THE
SMALL
COMPUTER
CONNECTION

"THE SMALL COMPUTER CONNECTION"

Telecommunications for the Home and Office

Neil L. Shapiro

A Micro Text/McGraw-Hill Copublication
New York, N.Y.

NOTICE TO READERS

The author welcomes your EMAIL through the Compuserve Information Service. Please address your comments or questions to Neil Shapiro—70001,1056.

Library of Congress Catalog Card Number: 82-062817

ISBN: 0-07-056412-4

Micro Text Publications, Inc.
McGraw - Hill Book Company
1221 Avenue of the Americas
New York, N.Y. 10020

Menus and prompts on The Source network are copyrighted by Readers Digest, Inc. and are reproduced here with permission. Menus and prompts on the Compuserve Information Service are copyrighted by the Compuserve Information Service, Inc. and are reproduced here with permission.

Apple II, Apple II-Plus, and Apple IIe are registered trademarks of Apple Computer Co.; Atari 800 is a registered trademark of Atari, Inc.; IBM is a registered trademark of International Business Machines Inc.; TRS-80 is a registered trademark of Tandy Corp.; Micromodem II is a registered trademark of Hayes Micro-computer Products, Inc.; ASCII Express—The Professional and Online are registered trademarks of of South-western Data Systems, Inc.; Videoterm is a registered trademark of VIDEX Corp.; Dan Paymar is a registered trademark of Dan Paymar Inc.; MX-80 and MX-100 are registered trademarks of Epson America, Inc.; VP-3501 is a registered trademark of Radio Corporation of America; The Source is a registered trademark of Readers Digest, Inc.; D-Cat is a registered trademark of Novation, Inc.; Electronic Bookshelf is a registered trademark of Mc-Graw Hill Book Co.; MicroNET and Compuserve are registered trademarks of Compuserve Information Services, Inc.; Dialog and Knowledge Index are registered trademarks of Lockheed, Inc., Dow Jones News/Retrieval Service is a registered trademark of Dow Jones Inc.; Visidex is a registered trademark of Visicorp, Inc.; The Data Reporter is a registered trademark of Synergistic Software; Robot 400 is a registered trademark of Robot Corporation. Rainbow 100 is a registered trademark of Digital Equipment Corp.

The data base services comparison table in Chapter Seven was printed by permission of Innovation magazine.

*To my wife, Irene, who sees me bent over
the computer for hours every day—and never
knows for sure if I'm working or playing.*

CONTENTS

Introduction

The new art of communicating through your personal computer opens a whole realm of intriguing and practical applications. This book details exactly how anyone — regardless of their level of computer experience — can join in this expanding world of computerized telecommunications.

Our system is based on the popular Apple II, Apple II-Plus and Apple IIe computers. At least one disk drive is required. Other systems, of course, will work equally well in telecommunications.

From the first chapter, which discusses how computers can communicate over everyday telephone lines via a device called a modem, to the final chapter which shows how to arrange and file telecommunicated information without need of paper or filing cabinets, this book has been written as a guide to exact methods and applications. There are no sociological predictions here—only those implied by the fact that we are able to show so many seemingly futuristic goals which can easily be accomplished today, by computer novices using equipment now available everywhere.

When you make the small computer connection, your personal computer will become a gateway to possibilities and procedures that change your life at home and in business.

We show how to access the world of the giant computer networks. How, once connected to national services, a user has a wide variety of interesting options for communication. From being able to get the most out of instantaneous and reliable electronic mail services, to finding one's way around the complexity of hundreds of informative areas, this book guides the newcomer every step of the way.

And, while the giant computer networks are of great interest and utility, the reader will also be introduced to the burgeoning, exciting realm of the Public Access Message systems (PAMS). These call-in systems are electronic bulletin-boards, and they run on over a thousand small and privately-owned computers throughout the country. Open to all callers, they cover everything from politics and medicine to adventure games and general conversation. Odds are that, in this book's extensive appendices, everyone will be able to find at least a few of these PAMS to call within their local telephone area. And, don't worry if a few long-distance calls seem tempting, because the appendices will also provide a table of complicated commands to reduce long-distance connect time.

The Hayes Micromodem II has two main parts. The circuit board plugs into the I/O slots on the Apple II's motherboard. The Micro-coupler connects the circuit board to the telephone line. The entire unit can be installed in less than five minutes.

Further, the reader will be able to take the seemingly giant step of setting up his or her own small computer network. Whether you would run a bulletin-board for the sheer fun of it, or establish a security-conscious, password-protected means of keeping the most remote of field reps in contact with the home office for your business—here is a true guidebook. We show exactly how to set up such a network in an inexpensive and simple manner. Every step of the way, from the original installation of the software to the ways computers and terminals can all interconnect via "protocol" uploads and downloads, the techniques are fully documented and called-out.

Through the use of what we term our "do-it-yourself system" we demonstrate exact methods to put all of our ideas into instant use. But, even those who choose to use other systems will find in this book a wealth of ideas and techniques. While our system is the easiest way to put into practice everything in this book—everyone will be able to use the majority of what we detail. Owners of the Atari 800, IBM PC, TRS-80 and other popular computers will be able to treat this book for most of the way as an ideabook on how to best utilize their own systems. And, in the instances when we do talk about our particular system, owners of other systems may take note of what each part of the system does and use the information as the basis of comparison shopping to equip their own setups.

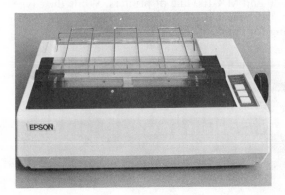

A good printer is a must. The Epson MX-80, shown here, is capable of printing graphics and text. The print quality is acceptable for all but the most formal applications.

For those who do want to follow along in a step-by-step manner, and who want to put together one of the most powerful and moderately-priced telecommunication systems, here is the equipment with which this book makes yesterday's dreams today's reality:

Our do-it-yourself system is based on the popular Apple II computer — either the Apple II, the Apple II + or the latest "Revision-E" Apple machines. At least one disk drive and controller card is required, and two drives are preferred but not vital. We add to the basic Apple computer various peripherals and software all from reputable and proven manufacturers.

The Hayes Micromodem II, a plug-in-circuit card for the Apple, is fully detailed in our first chapter. Now considered one of the standard modems in the field, it will allow you to communicate with all the services listed in this book, and to fully utilize all of the recommended software.

ASCII Express — The Professional is a terminal program from Southwestern Data Systems, and it too is fully discussed in the following chapters. We will show how this program will get the very most out of the above hardware.

Optional, but highly recommended, is the Videx Videoterm which is another plug-in circuit card for the Apple. This will convert the Apple's upper-case, 40-column display to 80 upper- and lower-case characters across the screen. It is fully supported by the *ASCII Express* software as well as by accessory software also available from Videx. But, if your budget quails at the additional expense of this 80-column card, our system can also be enhanced with at least lower case through a plug-in device called the Dan Paymar Lower Case Adapter.

All of the above software and hardware should be available at most computer stores.

To the main system you may want to add two further peripherals, a printer and a portable terminal. The printer will allow you to maintain hard copies of all your telecommunications. The portable terminal will allow you to keep on telecommunicating even in places you could never take your full computer system.

Today's selection of printers that will work with our recommended system is truly vast. You can get type quality just about indistinguishable from the print in this book. Or, you can get very inexpensive and "computer-looking" type styles. We like the route in between these methods and have tested, and found to be absolutely reliable, the Epson MX-80 and MX-100 line of printers. While their type styles, under close examination, will be seen to be computer-produced, their printouts are sufficiently eye-pleasing for the most formal of documents.

A portable terminal such as the RCA VP-3501 makes a handy accessory. This unit connects to any telephone and TV for true portability.

New portable terminals are being introduced every day. Without any computer power of their own, they allow you to connect to your computer and others from just about anywhere in the world there is a telephone. Our system has been reliably used with the RCA VP-3501 Videotex terminal. The VP-3501 can be quickly attached to almost any telephone and TV set combination. From a hotel room to a telephone booth, you can indeed take the Information Revolution with you.

We hope that you will find this book provides an intriguing entry into the ever-expanding and always exciting world of telecommunications.

1. You and Your Modem

Before you embark into the world of telecommunications you will have to have a modest understanding of a few of the components to which your computer is attached. The modem, and the serial port, are important accessories.

Your computer operates in a binary language of ones and zeroes. This binary language is due to the fact that all a computer really knows how to do is to throw thousands of microscopic, internal switches very quickly to off or on states.

For instance, when you type the letter "A" on your keyboard, the computer's brain has its own two-numbered way of handling that letter. It is called the ASCII binary code. In this binary system, the letter A is really the binary number 1000001. B is 1000010 and so on.

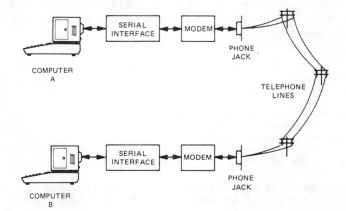

Telecommunications is the process of transmitting and receiving information, over the phone lines, by computer. As the text details, a modem and serial interface are needed at both ends of the connection.

Suppose you have word-processed a business letter, and have stored that letter on disk. What is really stored on disk is the binary encoded sequence that represents each of the letters one after the other—a very long string of ones and zeroes.

How can your computer send those binary numbers to another computer via a telephone connection? You must use something called a *modem* connected to a *serial output* which may be built either into the computer or the modem.

A modem simply converts the ones and zeroes to tones. These tones can be sent on the telephone. One tone means a number 1, the other tone means the number 0. These tones are sent very quickly—so quickly that if you listen with your ear it sounds much like a rainstorm, a hectic and hurried sound that only another computer could understand.

But before the modem can convert the 1's and 0's to tones, it has to receive them from your computer properly. To do this requires the serial output.

The 1's and 0's travel inside your computer as voltages on the computer's internal printed-circuit wiring, called a buss. There is a voltage level which means a 1, and a different voltage level which means a 0.

Remember that the letter A had seven 1's and 0's? Most computers can process separate pieces of information made up of eight 1's and 0's. The building blocks, the 1's and 0's are often called "bits" and the eight-bit pieces of information are called "bytes." The 7-bit ASCII code is prefaced with a leading 0 bit during a computer's operations. You do not ordinarily speak letter by letter, but in complete words. Your computer operates the same way and wants to speak in complete bytes.

Inside the computer there are eight "wires" that are laid out physically parallel in the computer's printed-circuit buss. When it wants to send the letter A to the disk or from the keyboard to the screen all eight bits of that byte (including a leading 0) can be sent at the same time. Each voltage representing one of the eight 1's and 0's goes on its own "wire" and all are sent at the same time.

Obviously this is far faster than sending a 0, then a 1, then a 0, then another 0, another 0, a 0 again, then another 0, then a 1—and so on.

But imagine the problems of eight people talking at once on a telephone. An audio circuit, which is only one line of communication, not eight, can only clearly handle one bit—or tone—at a time.

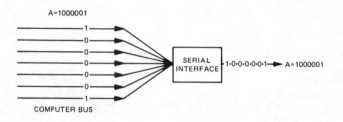

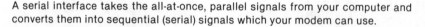

A serial interface takes the all-at-once, parallel signals from your computer and converts them into sequential (serial) signals which your modem can use.

So, a serial output takes this eight-bit-wide river of information from your computer. It then changes this parallel data into a narrower stream that is only one bit wide. The letter A is changed from the all-at-once, parallel, signals of 01000001 to an Indian-file serial output of 0, then 1, then 0, then 0, then 0, then 0, then 0, then the last 1. At this point, the leading zero is generally removed.

Your modem then takes each of these voltages meaning 1's and 0's following after each other and encodes them into those rainstorm-sounding tones. The voltages that mean 1's are encoded to a tone of 2,225 Hz (or vibrations per second), the voltages that mean 0 are encoded to tones of 2,025 Hz. These crystal-clear tones may then be sent from your computer onto the telephone line. This process is called "modulating" your computer's signal. There is another process—called "demodulating"—that will let your modem decode what the other computer is saying. Remember the tones that your modem is sending—2,225 and 2,025 Hz? The other computer is talking back down the telephone lines, also sending a medley of two tones carrying its own message of 1's and 0's. But these tones are slightly different notes. The other computer is sending tones of 1,270 Hz to mean a 1 and 1,070 Hz to mean a 0. Your modem catches these tones and converts them into voltages. The voltages are sent to the serial interface which this time re-arranges the serial Indian-file of bits into an 8-bit-wide stream again.

(HIGH AND LOW VOLTAGES)   (2,225 Hz = 1,2,025 Hz = 0)

Sequential, or serial, signals, arrive at your modem where they are encoded into tones. These audio tones are transmitted along the telephone lines.

Because your computer and the other computer are each sending different frequency tones both machines can be talking at once. Your modem will not be confused as to which is saying what. This process of modulating and demodulating is why the word modem was coined. As it turns out, the word modem is quite descriptive.

At this point, you now know enough about modems in general to skip ahead to the end of this chapter and install your modem in our do-it-yourself system by following our instructions and those supplied with your modem and computer. But, if you'd like to find out what such instructions as "Set the baud rate to 300" might mean technically, then read on.

PARITY

Transmitting a sequence of tones to represent a string of digits is a workable idea. But how does a modem recognize when one character is done being sent and the next tone-encoded character is beginning? The way most computers (all that you are likely to ever run into) handle this is by a method called asynchronous transmission. Some computers even add a process called parity checking. Like many processes in the world of computing, this one has a complex-sounding name for a logical and easy-to-comprehend process.

Each character in the ASCII 7-bit code has a *start bit* added to it. The start bit is a 0 and tells the receiving computer system to expect an ASCII letter. After the letter is sent a bit called a *parity check* may be tacked onto the transmitted character. In some systems, the parity check is the value 1 or 0 which, when added to the other bits in the number, will make the sum come out odd or even. Even parity-checking means that the bits in each character—summed with the parity bit — come out even. Odd parity-checking means that they come out an odd number. If you are in an even parity system and your computer system gets an odd sum during a parity check, then there has been a transmission or reception error.

The parity bit is not used too much in small computer telecommunications. But, *stop bits* are. Stop bits are one or two binary numbers which follow the parity check bit. The stop bits are end-of-letter "flags" and alert the receiving computer that "this is all for this letter, next should be the start bit of the following letter." Most of the computer systems we will be using will be requiring one stop bit.

BAUD RATES

When you drive your car, you measure your speed of travel in miles per hour. When you transmit words and information by computer your speed is measured in Baud. Unfortunately, a measurement such as 300 Baud does *not* translate as 300 words per minute. Nor does it even translate as 300 bits per second.

Remember that the signals which travel down the phone line each represent one bit—either a 1 or a 0. If each signal usually took a full second to send, it would take eight seconds to send the letter A (including one trailing stop bit). Obviously, the tones must be very short in length, because the shorter in length they are the quicker you will be able to communicate through them.

Baud rates are determined by mathematically taking the reciprocal of the time duration of the shortest signal element. Most computer-to-computer transmissions today take place at 300 Baud. In other words, each signal lasts a trifle over 3.3 milliseconds.

And, at 300 Baud, it is easy to calculate the number of characters sent per minute. The seven-bit ASCII code is prefaced by one start bit, suffixed with one stop bit. Each character has nine bits so 29.7 characters may be sent a second.

The only other Baud rates you are ever likely to run into are 1200 and 110 Baud. The 110 Baud is a slow rate that was once popular but is now being phased out and is almost completely unused. The 1200 Baud rate may be the coming thing. Right now, modems capable of this speed are expensive compared to 300 Baud modems. And, most information utilities charge more money for 1200 Baud connections. When prices come down, 1200 may be the wave of the future. But a Baud rate of 300 is today's standard rate of speed.

DUPLEX

Remember that we said two computers could be speaking at the same time, if they were both using different tones. This simultaneous, and most common, method of operating is called full duplex. If, however, the computer at either end of the circuit can only recognize one style of tones then half duplex must be used. In half duplex, your modem can only receive or transmit but cannot do both at once.

Full duplex is faster and is most often used. There are some tricky things, though, about duplex that you should be aware of. The most often encountered is the "double-letter" problem.

Full duplex is sometimes called echo-plex. The reason for this is that a computer set to receive full duplex sends out a character over the modem, then expects that the other computer's modem will echo that character back to it. This echoing of every letter is what you see on your video screen as you type. Type the letter A— and the letter A seen on your screen is the echo from the other computer.

If you are typing and all is well, you simply see the words you expect as you type.

But if you type "Hello" and you see it on your screen as "Headgfdfgd" then the echo is wrong and something in the connection is garbling the information.

Full-duplex is faster, as it allows both computers to talk at once. It is also more secure, as the echo is a continuous check on the quality of the connection.

If you are set for half duplex, your computer system does not expect an echo. As you type a letter it is simply routed once to the modem, and then automatically displayed on the screen. What you type is what you see, no matter what it is the other computer is receiving.

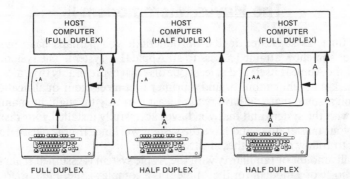

FULL DUPLEX   HALF DUPLEX   HALF DUPLEX

In full duplex operation you see on your screen what the other computer echoes back to you. In half duplex, you see what you are typing and there is no echo. If you are in half duplex and the other computer is in full you will see both what you type and the echo, or double letters.

Another problem may occur in the half-duplex mode. A second computer which expects full duplex will echo your letters back to you. You will be receiving the echo from the other computer as well as having your own system send letters to the screen. Typing "Hello" will result in "HHEELLLLOO." If you do see the double-letter effect it is almost positive proof that you are running the wrong duplex.

SUMMARY

Computers communicate over the telephone lines by using modems. A modem modulates the computer's outgoing signals into tones which can be sent down the telephone line. The modem receives and demodulates tones from the other computer's modem back into digital pulses the receiving computer can use.

A modem expects to send and receive signals one after the other in a serial fashion. Because of this, the computer or modem must be equipped with a serial port or serial interface. This serial port will convert the computer's parallel signals into serial signals to the modem. It will accept the serial modem's receptions and convert them back into parallel signals for the computer's internal wiring and logic scheme.

Letters and characters are sent under the conventions of ASCII code. They are sent with stop bits and start bits added onto the code. Each encoded letter may also contain a parity checking bit (but this last is seldom used in small computers).

Baud rate is the measure of how fast tones are being transmitted. The most usual rate of transmission is 300 Baud.

Full duplex means the computer your own computer is connected to is echoing what you are typing back to you. Half duplex does not involve an echo. Full duplex is faster and surer and is the most often-used method.

The Do-It-Yourself System
The Hayes Micromodem II

By this time, as suggested in the introduction, you perhaps have purchased, rented, or somehow attained access to an Apple II + system. The system is composed of the Apple II itself and at least one disk drive, preferably two, along with a good black-and-white monitor and a printer that meets your qualifications.

We will assume that you have read at least part way through the manuals that arrived with the system and that you have successfully installed your disk drives, booted your master disk, and have the system working. If not, please do not yet attempt to follow along in these instructions.

A small amount of familiarity with the Apple system is assumed from now on.

The choice of a modem for the Apple II is a complex issue. First, there are two types of modems. One style is called a direct-connect modem and the other is called acoustic coupling. An acoustic-coupled modem requires that you place your telephone's handset into a cradle which contains a microphone and receiver. The computer then telecommunicates by sending audio tones into the handset's microphone and picking them up from the handset's speaker.

All the acoustic coupling modems have a few drawbacks for the Apple and most other computers. Usually, it is necessary to already have, or to buy, a serial interface card for connection to the computer. Another drawback is that environmental noises can and almost always do cause some communications errors.

A direct-connection modem is, by and large, the better way to go. As it connects right into the telephone line room noises will never bother it. Some direct-connect modems have their own serial interfaces. Others require that there be one already installed in the computer.

Of the direct-connect modems for the Apple II, two of the most popular are the Hayes Micromodem II and the Novation D-Cat. Both even include a built-in serial port. The Novation offers a Baud rate of 1200 (but only in half duplex unless an additional circuit board is purchased) while the Hayes' top speed is 300 Baud.

We are going to build our system on the Apple II using the Hayes Micromodem II. This is because the Hayes modem has become almost a standard in the field.

The Hayes modem has two main parts: a circuit board which plugs into one of the slots on the Apple's main circuit board, and a smoked-lucite covered chassis called a microcoupler. A cable connects the microcoupler to the plugged-in circuit board. A supplied telephone cord leads from the microcoupler to the wall jack. This modem will work on either rotary- or pushbutton-dial phone service.

Before you get ready to install the circuit board in your Apple, make absolutely certain that the power is off—and is not going to surprise you by coming on when you least expect it to. The best thing to do is to unplug the AC cord from the back of your Apple.

Remove the top cover from your Apple by gripping it at the rear and carefully popping it off the velcro-style pads which hold it to the case. Do not allow the front of the cover to press or lever against the keyboard. You have probably done this before when you installed your disk drives.

The Hayes manual tells you that the Micromodem II will work in any of the Apple's plug-in slots except the leftmost one, slot 0. This is true, but by convention, slot #2 is where most software expects to find your modem. Because the Hayes manual's tutorials are written as if the modem were in slot #3, you will *temporarily* install it in slot #3. But, in everyday use, it will be in slot #2.

Follow the instructions in the manual to connect the cable which goes from the microcoupler to the circuit board. Now, all you have to do is to plug the telephone cable from the microcoupler into a modular wall jack—and to inform the phone company.

If you do not have a modular phone jack, but instead have one of the older varieties which have four round holes spaced about three-quarters of an inch apart you will have to replace the old jack with a new style one. You can call the phone company to do this for you, or you can quite easily take care of it yourself.

Many electronic stores, such as Radio Shack, sell various types of hardware to do this conversion yourself. The easiest way we know of is Radio Shack's modular phone-plug converter. This cube-shaped device simply plugs into your existing wall jack. On one of the cube's faces is a modular jack cutout.

Or, if you want to get a little fancier, the same stores should sell a variety of conversion jacks. If you don't mind some very simple rewiring you can replace your entire old jack with a new one. You might even convert to a flush-mounted wall style. These options are all fast to install and inexpensive—very inexpensive compared to the cost of an installation call.

Make certain that you do inform the phone company as the manual tells you to. The information about the FCC registration number (BI986H-62226-PC-E), the ringer equivalence number (0.4B), and the use of a modular phone jack (USOC-RJ11W or USOC-RJ11C) is all you should be required to tell your telephone company's business office.

Once you have the modem's circuit-board installed in your Apple, the microcoupler attached by its cable to that board, and the telephone cable from the microcoupler plugged into your modular phone jack—you have just joined the communications revolution.

The micromodem has a very reliable self-test feature. Use the Hayes-supplied software to run this self-test. If all is OK, proceed.

At this point you may run a few quick tests to make certain that your modem is working all right and to verify its self-test. First initialize (turn on) the micromodem by typing IN #3 and return if you are in either BASIC. (From the Apple's monitor mode with the asterisk prompt, you would have to type 3 control-K and return.)

From this point, whatever commands you issue to the micromodem must be preceded by a control-A. Hit control-A and you should see on your screen the words: "MICROMODEM II:?"

At this point, if you would like (and we do recommend it), you may follow the rest of the examples given in the Hayes manual. However, in our do-it-yourself system you will never be running the modem in the stand-alone style described in the manual.

The micromodem *is* capable of running without supporting software. The manual details how you can do this. But, software that uses the micromodem is now very plentiful, and the best programs are surprisingly inexpensive.

NON-HAYES NOTES

If you have opted for an acoustic-connected modem for your Apple, you may still follow along with all the rest of the do-it-yourself system. You will, however, have to set up your modem so that it will recognize and work with all of the systems and networks with which we will be interconnecting.

Most acoustic modems have switches with which you can set communication defaults. Set the parity switch to 0 or none. The systems that we will be calling do not use this parity-bit error checking. Most systems that involve microcomputers on one end do not.

Set the Baud rate switch to 300. This is the more-or-less accepted standard. You may find some systems that can use a faster rate (which some modems may support) but it is very unlikely that you will ever come across a system that is limited to only 110 Baud.

Set the Full/Half Duplex switch to Full Duplex. This echoing system of allowing both computers to communicate at the same time, along with continual checking of the connection, is the form of duplex you will most often encounter.

The stop-bit switch should be set at 1. One stop-bit, as detailed earlier, is the standard for a transmission speed of 300 Baud.

If you do use an acoustic coupled modem, a good tip is to replace the microphone in your telephone's handset with a condenser microphone. These special microphones are usually available at computer stores. You just unscrew your telephone's mouthpiece, allow the old carbon-granule microphone to fall out, drop in the new microphone and screw the mouthpiece back on.

The problem with the standard-equipment, carbon-granule microphone is that it is subject to "packing." During a normal conversation the telephone is being moved around and the carbon-granules in the microphone are being vibrated. But, in telecommunications, the handset lies without moving in the acoustic coupler, often for hours. The carbon-granules clump up and the reliability of the communications is destroyed.

If you have not duplicated our do-it-yourself system, you will still be able to make use of virtually every part of this book with the exception of the next chapter, which discusses Apple-specific software.

SUMMARY

You have now installed a modem in our system of the Apple II equipped with one or, preferably, two disk drives. If a Hayes micromodem is being used, you have checked it out and run its self-test feature. If an acoustic modem is employed, you have set the switches so as to correspond with the way you will be using the modem later on in this book.

You now have the hardware set up just about as you need it. The software—the terminal program—is what will tie all of your hardware together into one telecommunicating package. This software is the next thing you must consider.

2. Your Computer As a Terminal

In order to connect your personal computer to one of the networks explained throughout this book, it will become necessary to run a terminal program on your computer. A terminal program makes your computer act like or *emulate* a terminal. If you know what a terminal is then that explanation may be all you need for definition. But, likely enough, the word "terminal" needs some further explaining of its own.

A terminal is a device that allows you to connect to a computer system—to enter and to read out information. A terminal has a built-in modem or connects to an external one. It may have a typewriter-style keyboard, a keyboard with far more keys than on a typewriter, or a specialized entry system with only a very few keys. It will have a way of displaying information; either on a printout or on a video screen.

Most people deal with some sort of terminal almost every day. The airlines use terminals at their ticket counters to connect to their own computer network to assign seats on flights. If you go into a store and your credit card is placed into a small machine to be "verified"—that is a terminal. In this case the small terminal is connected (via a phone link) to the bank's computers. The terminal reads information automatically from an encoded strip on your card. Even many grocery and department stores have their own computer systems with every cash register being a terminal. In a store like this, the register-terminals not only handle direct money-changing but also send to the store's main computer a purchase-by-purchase inventory update.

Terminals are not created equal. There are dumb terminals and smart ones. A dumb terminal is simply a keyboard and display along with a modem connection. On receive—receive is all that it does. The information flows in, is displayed and is then lost. On transmit what you type is what you send. A dumb terminal is cheap but provides no shortcuts or cost-effective means of telecommunicating. You would probably not want to use a dumb terminal as your primary means of network access but—as we will see later in this book—dumb terminals can be very handy when you set up an extended network of your own. A smart terminal may have many different features depending on its electronic IQ—which is almost always directly relative to its cost. Let's take a look at a few of the most popular and necessary features on these smart terminals. These are the type of features a good terminal program should add to the capabilities of your personal computer's telecommunications system.

Complete Modem Compatibility: The terminal program should allow your computer to take full advantage of every one of your modem's functions and features. You should be able to operate in both full- and half-duplex, change to the various Baud rates supported by your modem, auto-dial if your modem is capable of that, and hang up and exit the terminal program. These are the basic features that you should look for no matter what your system or modem combination is.

Buffered Text: A dumb terminal accepts information, displays it, and the words just scroll off the top of the screen into oblivion. Most smart terminals, and all good personal computer terminal programs, allow you to store some incoming text and information in a "buffer." This buffer is computer memory. The more memory, the more text can be stored. Once the text is stored in memory it can be looked at later (on a display) and stored in a file written electronically to a floppy disk.

Editing: When you connect to most networks the time you are spending is, literally, money. If you are calling one of the information utilities such as The Source or Compuserve you are being billed directly for your "connect time." Even if you call a free service it is likely to be long-distance and you are paying for the call. A quick way to lose money is to try to write a long file while seated at the keyboard. An editor is a simple word-processing program that lets you type and store your information before you call the computer. Then you can quickly send the information ("upload") and avoid needless connect-time charges. The editor can also be used to clean up, change, or delete parts of the buffered text mentioned above either before or after it has been stored on a floppy disk. The editor function is one of a terminal's most important features. It should be a strong part of the terminal program you choose to run on your personal computer.

Keystroke Savers: A lot of things that you will be typing, such as passwords and account numbers can quickly get repetitive. Not only is it boring, but it is time-consuming—keep in mind those connect costs for every minute you are typ-

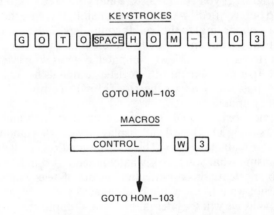

Macros allow you to type long entries with just a few keystrokes. They can save you typing repetitive commands or store and send commands you might forget.

ing. Many smart terminals offer user-defined keys. If your account number is AB456 you can have one key that, when pressed, sends the "character string" AB456. A terminal program copies this feature of smart terminals by using "macros." As there is no way to literally add more keys to most computers' keyboards a smart terminal program lets you define macros for your user-defined phrases. A macro is usually accessed by hitting a "prefix key." For instance, hitting just the A key will type an A. But, if it is macro-defined as such, hitting the prefix of control-W followed by A will type AB456. Macros save time, trouble, and are a great convenience if you have more than just a few passwords, account numbers, and such to remember.

A good terminal program is like a good, fast car. You may never need all that power and speed—but it can certainly be a help to have it available.

It does not pay to scrimp on buying a terminal program for your system. Get the very best you can find, even if you are not certain you will need all its features. The more you use such a system the more use you will find for all the "bells and whistles." Keep in mind that when you add the cost of the program to that of the computer and the modem and the printer, even an expensive program is only a fraction of the total system's cost.

SUMMARY

We have seen what a terminal is — a combination of keyboard, display, and modem capability. We have also looked at what makes one terminal dumb and another one smart, and have begun to find out what qualities of a smart terminal make it so useful in telecommunications. Some of these features include: the capability of accessing all of your modem's functions, the ability to save text in a buffer, the feature of being able to edit both the text you receive and the text you originate, and the user-defined keys called "macros." A terminal program allows your personal computer to "emulate" or copy many of the features (sometimes more) of a very expensive smart terminal.

You will want a terminal program that will make your computer into a very "smart" terminal. The terminal program we are about to describe for our system will turn your computer into a terminal with a very high "electronic IQ."

The Do-It-Yourself System
ASCII EXPRESS Professional Program

There are many and varied terminal programs for the Apple II. But the one which, as this book is written, offers the most power at the keyboard is *ASCII Express Professional* from Southwestern Data Systems (hereafter referred to as AE Pro). AE Pro performs just about every possible smart terminal function. It even adds some that no other programs or terminals allow.

Though it is a complex program, AE Pro's capabilities are structured so that their use is quite easy. The manual is must reading for anyone who wants to use all of AE Pro's powerful features. Yet, the on-screen help functions (called "menus") are detailed enough that you will find yourself learning how to use the program just from practicing with it.

AE Pro can be configured to almost any Apple II system using every Apple-compatible modem, printer, and serial interface that is available. This is done by running the "Install" program on the disk. The first time you boot the AE Pro disk, this "Install" program begins to run automatically.

Because our do-it-yourself system is fairly standard, there is little that you will have to do to install AE Pro.

AE PRO INSTALLATION

The first question the "Install" program will ask you is "Can you display lower case (Y/N)." If you have one of the lower case chips in your Apple as is discussed in the introductory chapter, answer "Y" for Yes. If you have elected to go with the optional Videx Videoterm 80-column board (as also detailed in the introduction) also answer "Y."

"Install" will then ask "Can You Display It Now? (Y/N)." If your system is set up in either of the above, recommended ways, answer "Y." From now on, AE Pro will display both upper and lower case letters. (You will only be answering "N" to the first question if you have neither an 80-column card nor a lower-case adapter. You only have to answer "N" to the second question if you have an 80-column card (not the Videx) which did not include a lower-case adapter.)

The next question lists ten various modem/serial interface combinations for the Apple. Type in choice 2 for the Hayes Micromodem II. To the following question, answer that it is in slot 2. (Remember that for best compatibility with various programs, the modem should be in slot 2, not slot 3 as suggested in the modem's manual.)

At this point, AE Pro will first print a message that it is installing your "communications driver" and will then take you into the main menu of the "Install" program. This menu will list all the many various and detailed variables that you can change in order to customize AE Pro to your own system and to your own operating preferences.

We will be returning to this menu later (see next chapter), but for now you have just about succeeded in setting up AE Pro on your system.

In fact, if you have the system we have detailed; your printer being the Epson MX-80 or MX-100 and Videx Videoterm for optional 80-column display, then you have done as much of the "Install" program as you need to do for now. Hit "S" for Save. The "Install" program will automatically recognize your printer and 80-column card and set AE Pro's printer and display defaults for you.

Only if you are using a "non-standard" printer or display device will you have to choose the "P" for printer and "L" as in display choices. At each choice level you will most likely be doing well simply by choosing the "O" or automatic setup choice.

After "Install" does its work you should see the "title" of:

ASCII EXPRESS "The Professional"
Version 3.46 © 1982 by
Southwestern Data Systems

on your computer's screen. (You may have a higher version number than shown above.) Just beneath that title you will see:

- >

and this " − > " is called the "system prompt." The prompt tells you two things
—that AE Pro is waiting for a command and that you are not presently connected
to another computer. A slightly different prompt

+ >

would show that you are connected to another computer and that AE Pro is awaiting your command.

If you are experienced in telecommunication you will want to follow the AE
Pro manual's suggestion at this point—which is to call your favorite system. But,
if you do not yet have a favorite system, and this is all new to you, let's take a few
minutes to first look at some of the AE Pro's "menus" of commands.

IT'S ON THE MENU

At the " − > " prompt, entering a question mark will take you to the first of AE
Pro's command menus. Hit the "?" key. If you have a 40-column display you will
see:

```
        ASCII EXPRESS "PRO" MENU 1
        ----------------------------
 !  = Display program status
 C  = Clear buffer
 D  = Dial or connect                ONLINE
 F  = Free buffer space
 G  = Get file from host (protocol)
 H  = Hang-up
 I  = Disk commands
 J  = View disk file
 L  = Load file to buffer
 M  = Macro select/review
 P  = Printer on-off                 OFF
 R  = Copy buffer on-off.            OFF
 S  = Send a file
 V  = View buffer
 W  = Write buffer to disk
 X  = Exit program
 -  = Display prefix characters
 2  = Display menu 2

+>_
```

"ASCII Express Professional" uses menus on-screen to help you learn and re-member its many, varied commands. This is Menu 1, which shows the most often used commands in this terminal program.

ASCII EXPRESS "PRO" MENU 1

!	= Display program status	
C	= Clear Buffer	
D	= Dial or connect	OFFLINE
F	= Free buffer space	
G	= Get file from host (protocol)	
H	= Hang-up	
I	= Disk Commands	
J	= View disk file	

```
L  =  Load file to buffer
M  =  Macro select/review
P  =  Printer on-off                              OFF
R  =  Copy buffer on-off                          OFF
S  =  Send a file
V  =  View buffer
W  =  Write buffer to disk
X  =  Exit buffer
2  =  Display menu 2
```

```
                    ASCII EXPRESS "PRO" MENU 1
                    --------------------------

 ! = Display program status       C = Clear buffer
 D = Dial or connect      ONLINE  F = Free buffer space
 G = Get file from host (protocol) H = Hang-up
 I = Disk commands                J = View disk file
 L = Load file to buffer          M = Macro select/review
 P = Printer on-off          OFF  R = Copy buffer on-off      OFF
 S = Send a file                  V = View buffer
 W = Write buffer to disk         X = Exit program
 - = Display prefix characters    2 = Display menu 2

+>
```

If you have the optional 80-column display, the above Main Menu will be split into two columns.

First, notice the commands with the word "buffer" in them. These are commands that you will be using frequently. Remember that a dumb terminal allows the incoming words to just scroll up the screen, roll off the top and vanish forever. But a smart terminal should have a way of capturing this information in a copy buffer so that you can look at it later. A really extra-smart terminal, like your computer running AE Pro, not only saves this information for later review, but will allow you many other options as well.

AE Pro's copy buffer can be turned on or off at any time. The R command toggles the buffer on and off. Hit "R." (If you are online, connected to another computer, you first have to get AE Pro's attention with a control-Q.) Now hit "?" to redisplay the Main Menu. Note how the "R-Line" now reads:

```
R  =  Copy buffer on-off                          ON
```

Hit "R" again and then again redisplay the menu with a "?". Now the line reads:

```
R  =  Copy buffer on-off                          OFF
```

just as it did when the program first came on. Using the same key to turn a function "on" or "off" is, as mentioned, called toggling. On this Main Menu, your printer may also be toggled either on or off by using the "P" command.

Remember, before any commands may be entered—you must first be seeing the " − > " or " + > " system prompts.

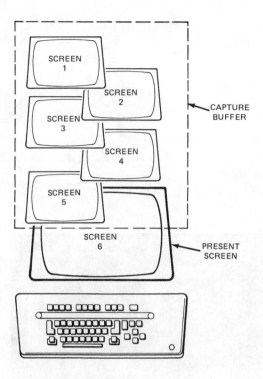

A capture buffer saves in memory information that has scrolled off the screen. In this manner you can retrieve previous screens of information. A capture buffer's contents can usually be edited and changed and then saved as a text file on your floppy disk.

Another word which shows up in this menu frequently is "file." This is short for text-file. Information is stored as a text-file written to your disk. The information is stored in the ASCII code mentioned in the first chapter. Words in a text-file form may be printed out, redisplayed on your screen, even changed or manipulated.

For now, hit "J" for "View disk file." The prompt will be overwritten by AE Pro's question "Filename?". Answer by typing Macro.Lib and hit return. On the screen you will see the words stored in that file. Later we discuss exactly what this Macro.Lib file is and how you can use it—but, for now, it is the only text-file on the disk that you can practice viewing.

You should now have the " – > " prompt again on-screen. Now hit "2" to bring AE Pro's second menu onto your screen. You will see:

ASCII EXPRESS "PRO" MENU 2

A	= Show control characters on-off	OFF
B	= Baud rate	300
E	= Echo duplex full/half	FULL
K	= Terminal chat mode on-off	OFF
N	= Set delay (transfer)	0
O	= Auto-Save on-off	ON
T	= Transpose H/RUB on-off	OFF
U	= Update/display macros	

Y	=	Editor	
Z	=	Screen format on-off	OFF
1	=	Display main menu	
:	=	Auto-disconnect on-off	ON
∧	=	Apple CAT port switch	INT
+	=	Auto answer (data)	
"	=	Keyclick on-off	OFF
/	=	Do CRC	
$	=	Emulation on-off	ON
#	=	Brief mode on-off	OFF
%	=	Run INSTALL program	
'	=	Answerback on-off	OFF

AE Pro's Menu 2 features some of the less-used but powerful commands which makes this program—and our equipment—one of the most powerful of telecommunicating systems.

```
ASCII EXPRESS "PRO" MENU 2
----------------------------------------
A = Show control characters        OFF
B = Baud rate                      300
E = Echo half-full                 FULL
K = Terminal chat on-off           OFF
N = Nulls - send delay
O = Auto save on-off               OFF
T = Transpose ^H/rub               OFF
U = Update/create macros
Y = Editor
Z = Screen format on-off           OFF
1 = Display main menu
: = Auto disconnect on-off         ON
^ = Apple-cat port switch          INT
+ = Auto answer (data)
" = Keyclick on-off                ON
$ = Emulation on-off               ON
# = Brief mode on-off              OFF
% = Run INSTALL program
' = Auto-answerback on-off         OFF
+>_
```

This secondary menu, while very important, will be used less often than the Main Menu. So, while there are certainly a lot of commands to AE Pro it is not really necessary to memorize them all. For one thing, you will only be using most of them infrequently. The ones that you will be using frequently are fairly mnemonic and easy to recall. Finally, if you ever should forget a needed command, the menus are easy to call onto the screen to guide you.

Let's take one quick look at just some of the features of the Secondary Menu.

The Baud Rate, and Duplex or Echo, ("B" and "E"), are already set to the most usual of defaults. Most systems, all of those that are detailed in this book, operate at 300 Baud and in full duplex. (There will only be a few times when you want half duplex—usually when connected to another personal computer which cannot echo back.)

The copy buffer can, as we said, be edited. The information saved in it can be manipulated—changed, partially deleted and/or added to before you save such information as a file on your disk. This is done by using the "Y" command to access the program's Editor. (We will get further into this, in detail, later in this book.)

AE Pro's Editor is the most advanced of any terminal program you are likely to see. For now, just be aware that AE Pro does indeed allow you to change—to word process— any information that you receive from another computer. Or, it will allow you to originate files for your computer to send to another computer.

From this menu you can also go back to the Main Menu, at any time, by entering a "1" at the " − > " prompt.

Notice how there is a choice on the Secondary Menu:

U = Update/display macros

and now switch to the Main Menu (enter "1") and note that the Main Menu has a command-choice which reads:

M = Macro select/review

These macros will, in the long run, save you much typing. They are one of the handiest and most surprising features of this terminal program. Later on we present a step-by-step method of how you use these two menus together to make your own set of macros custom-designed to your own telecommunicating needs. For now, just bear in mind that AE Pro does have this macro capability.

For the first few times that you connect to another computer or to a computer network it is actually better if you do not use the macros. The best way to learn your way around a new system, or to learn telecommunications in general, is to simply answer a system's questions or supply the system with commands typed in right from your keyboard. Using macros now would be a little like using the cruise control on your car before learning to control its speed with the gas pedal.

AE Pro also has one quite handy feature for touch-typists available at this menu level. The Apple keyboard has a different tactile feedback from your usual electric typewriter and far different from a manual typewriter. And, it has almost no audio feedback at all.

Note the line which offers:

" = Keyclick on-off   OFF

and enter the """" at the " − > " prompt. If you are like most touch typists the soft, but ergonomically reassuring clicks of the keys will allow you to type at full-speed just as if you were on your usual typing machine.

The many other commands in the Secondary Menu come in handy in various special instances, many of which will be detailed at various points during the course of this book.

SUMMARY

You have now installed the AE Pro terminal program on your own system. The menu-driven features of the AE Pro system have been detailed and you now know how to get a menu onto the screen and how to enter a command. The concept of "toggling" commands has been introduced. A number of AE Pro's powerful capabilities have been shown. Specifically commands relating to the copy-buffer, to macros, to the AE Pro editor, and to both menus have been given enough introduction to allow examination by use in the following chapter.

The next chapter will let you use AE Pro on-line, connected to another computer. For many, this will be the first phone call to the wide world of telecommunications.

3. Your First Phone Call

Now that you have your computer, modem and terminal software all working to-gether you are ready to enter the world of telecommunication. For your first phone call, look over the Appendix which lists the popular Public Access Message Systems (PAMS).

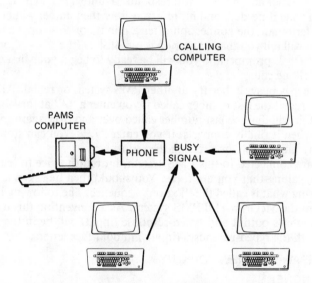

A Public Access Message System (PAMS) runs on a small computer. Other computers can call this system on the regular telephone lines. Only one computer can be connected to a PAMS at a time, other callers get a busy signal.

A cursory look at the Appendix shows a number of various types of PAMS. There are RCP/M (Remote CP/M), ABBS (Apple Bulletin Board Systems), Bullet-80 boards and many more styles. For purposes of this chapter, pick one of the PMS style (People's Message System) of boards to call so that you may follow along exactly with our step-by-step instructions.

Once you have familiarized yourself with one system you will be able to use all of them—particularly using the various downloaded information files also found in the Appendix.

Choose a PMS system that is nearby in terms of area code and cost to telephone. You will be billed by your phone company at the usual rates for a call to that area code.

If you are in the New York City area, or if you don't mind the long-distance charges, call with us "The Electronic Bookshelf," a PMS run by Mc-Graw Hill in Manhattan. The number is (212) 997-2488. Hours are 5 p.m.-8 a.m. EST Monday-Friday; 24 hours on weekends and holidays.

Like all of the PAMS, also referred to as "bulletin-boards," this system is a small computer network, as small as a network can be. There are two computers involved, yours and the one which is running the bulletin-board program. If you are using the Apple II with the AE Pro software and Hayes Micromodem II calling is a very simple process. Just turn on the computer with the AE Pro disk in drive one.

If you did the "Install" program properly as discussed in the last chapter (and the AE Pro manual), you should see the AE Pro title screen and the arrow-shaped prompt. (If you have not yet done the "Install" program, go back and do so.)

Dial the number as your software and system require. For our system, simply hit the "D" for Dial and AE Pro will respond "Number?". Type in the phone number (area code if needed) and hit return. Either the number will be busy and AE Pro will terminate the connection after a minute or so — or the screen will clear, AE Pro will tell you the connection is established (the " –> " prompt will change to a " + > " prompt) and you will be ready to begin your first computer-to-computer phone call.

If the number is busy, either try another PMS system or redial. AE Pro will automatically redial the last number called if you enter a "/" at the "Number?" prompt. And, it will dial the last number called over and over again (ringing the Apple's bell when it finally connects) if you enter "//". You can stop dialing a number at any time simply by pressing the return key.

Once you are connected to this PMS system, hit return twice to tell the computer you are calling that you are there. You should then be asked a few short questions during what is called the "log-on" sequence. The following is a log-on sequence from the McGraw-Hill PMS system. As a convention throughout this book, a computer's output and typed-response (input) will be in the following type styles, with the typed-responses (inputs) in bold-face letters.

PMS — MCGRAW-HILL, NEW YORK, NY

USERID (N = NONE) :**N**

YOUR FULL NAME ?**NEIL SHAPIRO**

CITY, ST. ?**BETHPAGE, NY**

PHONE NUMBER ?**212-262-4821**

YOU ARE NEIL SHAPIRO FROM BETHPAGE, NY

IS THAT RIGHT ?**Y**

CAN YOU RECEIVE LOWER CASE ?**Y**

Logging caller #8058 at 11:55 Eastern

Type "N" for system news.

Type "?" for commands.

McGraw-Hill Electronic Bookshelf (TM)

Featuring the latest on
Computers and Electronics

---type F to get started---

Command?

The above is a typical PMS log-on sequence and it is also very representative of the other systems as well, all of which require almost identical log-on information from the caller. Once you have supplied your name, phone number and address the computer files this information to disk. In your next call, you will only have to identify yourself by name and the computer will display your address from its files.

The "Command?" line is your means of utilizing the bulletin-board system. With a little practice you will soon be able to send and receive messages. Most of the commands are one-letter abbreviations. Besides commands there are also control-codes which allow you to manipulate the way the board is sending you information.

The first thing you should do on a new system is to find the help-files. This is usually done by typing "H" or a similar command. A typical PMS help file, just as you would see it on the screen, is in the Appendices. Also there are the help files for most of the other popular systems. Use our help file Appendix to study the command structure whenever you call a new system or want to brush up before calling an old one. One thing is for certain, it makes more sense to study such a list while you are off-line than to wait until you are connected to the system. Since it is usually harder to read a long file on a video screen than in a printed version, later chapters will detail how you can "capture" such a listing to print out at your leisure. (For now, refer to the help file Appendix.)

The first command you might wish to enter is "S" for Scan. This command allows you to scan only the "headers" of the messages that are presently on the bulletin-board. The message header tells you just enough about the message—its number on the board, the date and time it was posted, the subject, who sent it and who the intended recipient(s) are—that you can decide if you want the complete message.

A typical board may contain two hundred or more messages. Scrolling through all of them and remembering which to read can be tedious—so it is best to "mark" the ones that interest you for later recall. As discussed further in the help file, this is done by hitting the letter "R" during the header *which follows* the message you want to read.

Entering the "S" command will start the headers scrolling up your screen. Here is a typical example of the beginning of a typical board's scannable headers—this example having come from the McGraw-Hill PMS:

Start at msg# (2154/2692) ?**2154**

Msg# 2154 on 06/05/82 @ 19:38 (12)
Subj: DISK DRIVES FOR SALE, To: ALL
From: SERGE SRETSCHINSKY

Msg# 2155 on 06/05/82 @ 22:19 (12)
Subj: Forth, etc., To: Robert Shiller
From: mike cohen

Msg# 2158 on 06/06/82 @ 14:02 (12)
Subj: AUTOMATIC PHONE FOR SALE, To: ALL
From: GABE WIENER

Msg# 2159 on 06/06/82 @ 15:01 (9)
Subj: PRINTER FOR SALE, To: ALL
From: ALBERT BERG

Msg# 2160 on 06/06/82 @ 17:42 (18)
Subj: FREE SOFTWARE, To: ALL ATARI USERS
From: ronald luks

Msg# 2168 on 06/07/82 @ 09:07 (13)
Subj: copying, To: all
From: Sysop

Msg# 2171 on 06/07/82 @ 18:21 (6)   Pvt
Subj: NCC MSG, To: DON FRUEHLING
From: JOHN L. DIESEM
OK   2168

Msg# 2173 on 06/07/82 @ 19:19 (11)
Subj: DATA BASE:STOCKS, To: ALL
From: TED YOUNG

Look at the last line of message #2171. It reads "**OK**   2168   " and what it means is that the PMS system recognized that I hit "R" for retrieve in order to mark the previous message—#2168. In this manner, I "flagged" message #2168 for later retrieval. Message #2168 was from "Sysop" and was addressed to all. Sysop is a word you run into frequently; it means System Operator and refers to the person who runs and/or owns the bulletin-board computer to which you are connected.

Hitting return a few times would bring us back to the "Command?" prompt. From there, we can enter "*" which means to retrieve the messages marked during the scan. Such as:

Command ?*

Msg# 2168 on 06/07/82 @ 09:07 (13)
Subj: copying, To: all
From: Sysop

TO ALL BOARD USERS:
THERE HAVE BEEN A NUMBER OF INQUIRIES OF LATE REGARDING THIS
BOARD'S POLICY ON THE COPYING OF COPYRIGHTED SOFTWARE. THIS
PMS IS OWNED AND OPERATED BY MCGRAW HILL BOOK CO., THE
COMPANY DOES NOT TOLERATE ANY INFRINGEMENT OF THE COPYRIGHT
LAWS. EFFECTIVE IMMEDIATELY ANY MESSAGES ON THE SYSTEM
ADVERTISING THE COPYING, SALE, TRADING, OR OTHER ILLEGAL
DISTRIBUTION OF PROTECTED SOFTWARE WILL BE DELETED. THANK YOU.
    (sysop)

End msg #2168

The next thing you might wish to do is to leave a message on the system intro-
ducing yourself or your organization to anyone else who calls in. The message can
be typed directly from your keyboard (although later in this book we will show
ways to prepare such letters while off-line to send later).

Use the "E" for Enter Message command which takes you into the PMS message editor. You may type lines of up to 40 or 64 characters depending on how you
have your options set (default is 64). If you reach the end of the line you will not be
able to enter any more on that line. If in the middle of a word you will have to use
the back-arrow key to backspace. Hitting return works much the same as a carriage return on a typewriter, it advances you to the next line to be entered. Each
line will be preceded by a number so that you may keep track of your message's
length (many systems impose a maximum limit, often at 16 lines). A question-
mark prompt (?) will show you whenever the system is waiting for you to type in
some information.

Here is what the screen looked like as I entered a message into the McGraw-Hill
PMS:

Command ?**E**

To ?**All Apple People**
Subject ?**MAUG**

C/r to end.

1
?**The MicroNET Apple Users Group (MAUG) is one of the special**
2
?**interest groups offered on the Compuserve network. It is a**
3
?**very active bulletin-board with more than a thousand**
4
?**members from every part of the country. Many of the premier**
5
?**Apple programmers (such as Silas Warner and**
6
?**Mitch Kapor among others) are members as well as those who**
7
?**have just gotten into the world of Apple. Other than the**
8

?time-share charges from Compuserve, membership is free to all.
9
?Once you are a member you have access to dozens of programs
10
?for downloading as well as to the "CO" or conference line.
11
?In CO you can talk directly to other members — or join in on
12
?our weekly Sunday night meetings. You can now pick up
13
?the Compuserve package at most computer stores. Hope to see
14
?you on MAUG!
15
?Neil Shapiro
16
?MAUG Sysop
17
?

At that point, line 17, I just hit return rather than entering any more information so the system knew that the letter was finished. The system then sent the following prompt line (which I answered with a question mark):

(A,C,D,E,H,I,L,S,W,?) ??

Entering the question mark caused a sub-menu to come onto the screen that showed me what that abbreviated prompt-line was expecting in the way of one-letter commands.

-----   Message entry commands

A = Abort message. Return with Z.
C = Continue entry.
D = Delete line (Dx)
E = Edit line (Ex)
H = Help with message entry functions.
I   = Insert a line (Ix)
L = List lines from specified # (Lx)
S = Save completed message to disk.
W = reWrite an old message.

I responded:

(A,C,D,E,H,I,L,S,W,?) ?L10

And, the system responded by listing my message from line 10 onwards:

10
for downloading as well as to the "CO" or conference line.
11
In CO you can talk directly to other members — or join in on
12

our weekly Sunday night meetings. You can now pick up
13
the Compuserve package at most computer stores. Hope to see
14
you on MAUG!
15
Neil Shapiro
16
MAUG Sysop

I decided that I would enter, as part of my signature, my MicroNET account number (you will, of course, be reading more about MicroNET and such in a later chapter). To do this, I decided that I would edit line 15.

(A,C,D,E,H,I,L,S,W,&) ?**e15**

line #15:

Neil Shapiro

edit, enter new line or c/r to leave

?**Neil Shapiro, 70001,1056**
line #15:

Neil Shapiro, 7001,1056

edit, enter new line or c/r to leave

?

Hitting return presented me again with the abbreviated menu. This time I entered "S" for Send.

(A,C,D,E,H,I,L,S,W,?) ?**S**

The message was then saved to disk and became a part of the bulletin-board. The system said:

Msg# 2779. . . . saved
Password:**test**

The password (I entered "test") was so that only I (or the Sysop) would be able to delete that message. Then I logged off by typing "G" for Goodbye at "Command?" and went through the quick log-out procedure.

Command ?**g**

Comments/suggestions (y/n/r) ?**n**

Connect time 36:24

It's 12:31:28   Goodbye.

If I had wanted to, I could have left a comment or suggestion that could only have been read by the "Electronic Bookshelf's" Sysop. The system finally displayed how many minutes I had been connected, the present time of day in the board's time zone, and with a brief "Goodbye" broke the connection.

Of course you may—and should—want to spend a little more time on the board experimenting with the various commands. Not only is this new form of communication fun and exciting, but the skills you will pick up from practicing on one of these small, two-computer networks, these bulletin-board systems, will stand you in good stead later. It will be that much easier when, later in this book, we connect to the giant computer networks, known as the information utilities, and if you decide to follow our instructions for setting up your own extended networks for your own personal or business use.

4. The Information Utilities

While many of the private bulletin-board services introduced in the last chapter and further discussed in Appendix One contain literally reams of information, as well as the opportunity for personal enjoyment, they have a number of drawbacks for businesses. The nature of the PAMS (Public Access Message Systems) is such that they can best be used for recreation. They cannot be made to work for you. (Unless you set up one of your own . . . . But, more on that possibility later.) The PAMS are certainly one of the most enjoyable parts of the telecommunications *hobby*—and when you have honed your efforts and practiced on them the next step is to join one or both of the giant information utilities.

TIME-SHARING UTILITIES

The two major information utilities are the Compuserve Information Service and The Source. They have many similarities.

The Source and Compuserve are the first of the information utilities that are affordable, and powerful enough, to interest the small-computer telecommunicator. Both of these information utilities are large networks which run on very powerful mainframe computers. Because they are large networks, hundreds of people may be connected to the utility's databases at one time. But, through a process known as time-sharing, each user (unless he chooses otherwise) can have the impression that he alone is accessing the utility.

The big information utilities, for most of the country, are as close as a local phone call away. It does not matter that, for instance, Compuserve's eleven mainframe computers are tucked away in an air-conditioned room in Columbus, Ohio. From California or New York City, from just about every major U.S. city and from parts of Canada, all you have to do is to connect your computer via a local telephone call. While the user only "sees" the utility's computers, it is good to keep in mind that many computers along the way are being used to route data back and forth.

The mainframe computers are powerful enough that they can split up their time among the callers so quickly that each caller seldom has to wait for the computer to respond to his commands. The mainframe quickly takes everyone's commands in a roundtable sort of order—processing a little bit here and a little more there— and the net effect (when the system is working properly) is that everyone feels they are connected to their "own" mainframe computer.

Of course, like everything else in this world, sometimes the best-laid plans of men and computers both can go temporarily wrong. Many times there are just so many people connected to these utilities that the delays become more noticeable. Once in a very great while they can become annoying. And, at least once, you will probably be witness to a system "crash."

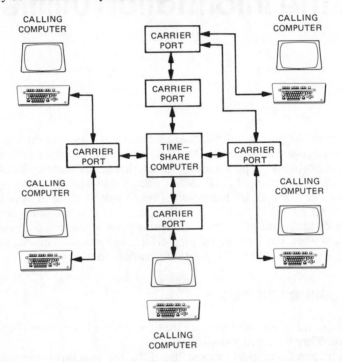

Time-share computers allow for many calling computers to be connected at the same time. Each caller connects with its most local carrier, then information is routed back and forth to the main, time-share computer.

But the most important thing to remember while on an information utility is this: If you hit "return" or enter any other command and nothing seems to happen—wait. It just may be that the load on the network is such that even the very powerful mainframe computers cannot "keep up."

All of this is outweighed, however, by the absolutely tremendous advantage of having such a great amount of information which may be accessed by so many people during the course of the day. Later, we will see how such things as electronic mail take advantage of an information utility's audience base and of the possible interactions between users who are connected to the mainframe at the same time.

YOUR TELECOMMUNICATIONS IDENTITY

Both of the major utilities will require you to enter your account number and your password. The first becomes your electronic "name" while connected to the system. While there may be other users who have everyday names similar or even identical to yours, your account number guarantees you a telecommunications identity all your own.

This does not mean that you will be giving up the use of your name on the utility. But you may have to remember that the salesman, George, in your third district is 70000,876 while Mary, your boss, is 70000,872. Most times the systems are structured so that if you are replying to a message or have already contacted a person you will not need to recall his or her account number, but it is safest to begin keeping lists. (Later we will show how to keep a lot of information right on the utility itself.)

Your password verifies to the system that the person using your account number really is you.

Temporary passwords are often assigned at the same time you get your account number. Then, as soon as you sign onto the system, you should change the password to one that you make up. Your password should be an unguessable combination of letters, punctuation and numbers — but make certain that you can remember it yourself. Once you change your password it is very difficult, and sometimes expensive, to recover from a lapse in memory. Even the utilities cannot ordinarily determine what your password is (although they can, if given permission and for a fee, usually recover files that would otherwise be lost if a password is forgotten).

Your password is the only verification you can offer the system as to who you are. Signing in with your account number is like printing your name on the back of a check, giving the password is like writing your signature. Never give your password to anyone.

The first dozen or so times that you sign-on to a utility, it is moderately interesting to type in your account number and password. But, this can quickly become a chore. Set up your macros so that one keystroke gives you your account number and another keystroke produces your password. (At the end of the chapter we will detail how to set up our do-it-yourself system to achieve this.)

MENUS AND COMMANDS

Once you are signed on to the information utilities, you will see there are two styles in which they can present their information to you. There is usually a menu mode and a command mode available to the user. When you first start out, you will likely be depending for much of the time on the service's menu structure.

A menu on an information utility is pretty much the same, conceptually, as a menu in a restaurant. It is a list of items available which you can "order up" by number. Most restaurant menus have a few major categories, some minor categories, and the dishes themselves are specifically detailed. Major categories might be broken down into Meat and Fish. The Fish section could be further subdivided into Shellfish and other categories before you actually find the name of the dish that you want.

The menus on an information utility work in much the same manner, but they are far more complete and the level of categorization is deeper than even the world's largest restaurant's menu might be.

The first menu usually lists around a dozen or so major areas of interest. Each area of interest may be composed of many separate databases. Each menu leads to another and more specific menu until you have finally come to the item of information which you want to "consume."

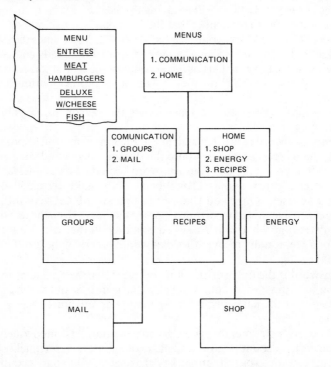

Menus in a restaurant describe the offered cuisine in categories—moving from general to specific. Menus on the information utilities do likewise for the services they offer.

While menus are easy to use, these video-style menus have one drawback that on-paper menus do not have. It is possible to spend more money on charges for reading the menus than on the tidbits you came in for in the first place.

Once you are connected to an information utility you are being charged by that utility at a set rate per hour. It doesn't matter (usually) what you are doing that hour. You could be deep within a database of stock prices, you could be doing any of a hundred useful things, or you could be reading menus and becoming frantic going from level to level trying to find your way to information you need—all the time you are connected is still being billed to your account.

Menus are a good way to be introduced to an information utility's structure. They are not the best way to continue using that utility—especially if you are often choosing the same databases or areas of the utility in which to spend your time.

Rather than wasting time unnecessarily in the menu structure, it is important to become more proficient at using the command mode. Command Mode is the preferred means of operating when connected to any time-sharing, hourly billing system.

Some forms of Command Mode give no menus, but only a prompt. The prompt in the Source is a " -> " and on Compuserve the letters "OK". The prompt means the system expects to receive a command. Various commands can take you directly to individual database areas and so bypass all of the menus. Obviously, you have to know what these commands are.

The Compuserve Information Service allows a form of Command Mode called "paging." On CIS many of the databases and services are assigned a "page." The page number appears on-screen while you are on that service. At the end of a menu there is a choice of options—including typing in "go XXX-XX" where the "go" command includes the abbreviated name of a service and its page number. (For example, "Go EMA-4" would mean go to the fourth page of electronic mail which happens to be the page that you go to in order to send someone a message.)

During the course of your telecommunications experience we will have opportunity to use all these forms of accessing information. We will be working with menus, with pages, and with advanced command modes.

Needless to say, the first call to an information utility can be a frustrating experience. It is very similar to learning a new language by being parachuted into a foreign country. For this reason, the information utilities offer varying amounts of free time for the new user to become accustomed to things.

Charges for each information utility vary. The basic charges are $100 one-time for The Source plus a connect fee of $7.75 per hour between 6 pm and midnight. Compuserve is cheaper in that there is only a less-than $30 one-time fee (which includes some useful software) and the connect charges are $5.25 per hour. Both utilities may be "purchased" from most neighborhood computer stores.

The remainder of this chapter will be a quick tour through a few of Compuserve's "high points" as a way of detailing some things you can expect in either utility.

A FIRST UTILITY CALL

Let's now examine how these menu structures would work during actual telecommunications. If you already have an account on one of the information utilities, you may follow along at your computer.

First, dial and connect with the carrier service. (With our AE Pro system the command is "D" for Dial.) Once connected, you will have to provide certain information so as to get signed-on to the information utility. Such a dialogue with Compuserve goes like this:

After connection is established, hit return twice. The following will result:

Host Name: **CIS**

The carrier service asks which host or mainframe computer you are connecting with. "CIS" indicates you want the Compuserve Information Service. When connected to CIS, you may begin to log onto the actual information utility:

User ID: **70001,1056**
Password:

Note that while your account number is visible as it is typed, your password stays invisible. The information utility computer—the host computer—does not echo that back to your terminal. Since the mode in use is full-duplex, and therefore depends on the echo for its display on the screen, you will not see what is being typed.

Once the utility matches your account number and password, and so assures itself of your identity, the first menu screen is presented to you:

CompuServe Information Service

1 Home Services
2 Business & Financial
3 Personal Computing
4 Services for Professionals
5 User Information
6 Index
Enter your selection number,
or H for more information.

!1

We will start here by entering a 1, to choose the category of Home Services. (As always, a computer's typing or output is in the regular typeface, a person's typing, or input, is indicated by the bold typeface.) Enter the "1" and the next screen displays:

CompuServe   Page HOM-1

HOME SERVICES

1 Newspapers
2 Weather
3 Reference Library
4 Communications
5 Shop/Bank at Home
6 Groups and Clubs
7 Games and Entertainment
8 Education
9 Home Management

Last menu page. Key digit
or M for previous menu.

!9

By choosing "9" to go to the Home Management subcategory, we are presented with:

CompuServe   Page HOM-80

HOME MANAGEMENT

1 Balance Your Checkbook

2 Calculate Your Next Raise

3 Calculate Your Net Worth

4 Amortize A Loan

Last menu page. Key digit
or M for previous menu.

But, now, let's say that we have suddenly decided against the home management subcategory. Notice that at the top right of each screen is a three-letter abbreviation with a hyphenated number. This combination is referred to as a page number. Look back and note the page number for the Home Services category was HOM-1. Let's get back to there. At the "!" we will enter the page number and a "Go" command:

!GO HOM-1

Sure enough, this takes us back to the Home Services category menu:

CompuServe   Page HOM-1

HOME SERVICES

1 Newspapers
2 Weather
3 Reference Library
4 Communications
5 Shop/Bank at Home
6 Groups and Clubs
7 Games and Entertainment
8 Education
9 Home Management

Last menu page. Key digit
or M for previous menu.

!3

This time we have chosen subcategory 3 "Reference Library." The next screen we see is another menu:

CompuServe   Page HOM-20

HOME MANAGEMENT

1 Movie Reviews
2 Popular Science
3 U.S. Government Publications

4 Video Information
5 The Victory Garden
6 Pan American Travel Guide
7 Gandolf's Report

Last menu page. Key digit
or M for previous menu.

!3

This time, after choosing the U.S. Government Publications category we do not go directly to another menu. Instead, we get this message:

Request Recorded.
One Moment, Please

At this point the information utility is transferring us from the simple menu structure directly into one of the many databases. In this case, the database we have chosen is the U.S. Government Publications database. We are waiting as the control passes from the computer that stores the menu structure to the computer which is storing the database. The wait is usually less than a minute, and the screen clears to show:

Thank You for Waiting

CompuServe   Page GPO-1

UNITED STATES GOVERNMENT
        PUBLICATIONS

1 Personal Finance
2 Health and Fitness
3 Automotive Topics
4 Food Preparation and Storage
5 Parent and Child
6 Energy Conservation
7 Consumer Notes

Last menu page. Key digit
or M for previous menu.

!3

Notice how detailed the menus are becoming as we continue to "zero in" on the information we need. We enter the "3" in order to see what type of U.S. Government Publications deal with Automotive matters and we see:

CompuServe   Page GPO-6

AUTOMOTIVE

1 How to Deal with Motor Vehicle

Emergencies
2 How to Save Gas and Money
3 Common Sense in Buying a Safe
  Used Car
4 Car Care and Service

Last menu page. Key digit
or M for previous menu.

!4

Checking for Car Care and Service sub-subcategory we finally come to the most basic menu, a table of contents. At this point every choice is an actual article or service. We are now viewing titles rather than categories.

CompuServe   Page GPO-1268

CAR CARE AND SERVICE
(Table of Contents)

1 Introduction
2 Accepting a New Car
3 The Break-In Period
4 Regular Care and Preventive
  Service Signals
5 Overheating Countermeasures
6 Checking Gasoline Mileage
7 How to Get the Best Service

Last menu page. Key digit
or M for previous menu.

Note that the page number for this table of contents is GPO-1268. Write it down or remember it. Now, if we wanted to, we could either retrieve particular articles or get back to the beginning of all this by using the "M" command (detailed below) which would—menu-by-menu—take us on a reverse trip. Instead, at the prompt we will enter:

!GO CIS-1

And that will take us back to the first menu page:

CompuServe   Page CIS-1

CompuServe Information Service

1 Home Services
2 Business & Financial
3 Personal Computing
4 Services for Professionals
5 User Information
6 Index

Enter your selection number,
or H for more information.

If at this prompt we should enter GO GPO-1268 page number we could skip all of those intervening menus again to get directly to the table of contents concerning Car Care and Service.

At this point you might want to look through the CIS Index which tells you the page number of literally hundreds of areas. You can get there by entering "Go IND-1" at any menu prompt. It is a good idea to print this index out as soon as you can. (If you have our AE Pro system the "P" command will print it out for you.)

Many commands are available at the menus. Hitting return at the menu prompt displays the following list of commands:

CompuServe   Page CIS-162

Brief Command Summary

* *

T — TOP menu page
M — previous MENU
F — FORWARD a page
B — BACK a page
H — HELP
R — RESEND a page
S n — SCROLL from item "n"
G n — GO directly to page "n"
N — display NEXT menu item
P — display PREVIOUS menu item

A little bit of practice in this menu area—coupled with some investigation of the Index before getting on for a day's telecommunications—will help you speed up your information access and cut down the costs of connect time.

The Source also uses menus but they are implemented in a slightly different manner from Compuserve's. Upon signing onto the Source the first menu is displayed:

WELCOME TO THE SOURCE   15:32

1 OVERVIEW OF THE SOURCE
2 INSTRUCTIONS
3 THE SOURCE MENU
4 COMMAND LEVEL
5 TODAY

Enter item number   3

Choosing the Source Menu sets up the Menu Mode:

THE SOURCE MAIN MENU

1 NEWS AND REFERENCE RESOURCES

2 BUSINESS/FINANCIAL MARKETS
3 CATALOGUE SHOPPING
4 HOME AND LEISURE
5 EDUCATION AND CAREER
6 MAIL AND COMMUNICATIONS
7 CREATING AND COMPUTING
8 SOURCE*PLUS

Enter item number or HELP **1**

From the main menu, let's just page through until we get to the first of the actual databases:

NEWS & REFERENCE RESOURCES

1 NEWS AND SPORTS
2 TRAVEL AND DINING
3 GOVERNMENT AND POLITICS
4 CONSUMER INFORMATION
5 SCIENCE AND TECHNOLOGY

Enter item number or HELP **5**

SCIENCE AND TECHNOLOGY

1 ELECTRICAL ENGINEERING
2 MECHANICAL ENGINEERING
3 SIMULATION
4 GEOGRAPHY
5 STATISTICS

Enter item number or HELP **4**

At this point you will see the following message — and one of the first differences between The Source and Compuserve becomes apparent:

THESE PROGRAMS CAN BE ACCESSED WITH THE INFO COMMAND FOR A
DESCRIPTION AND THE R COMMAND TO BEGIN THE PROGRAM.

GIVEN MONTHLY DATA ON TEMPERATURE AND RAINFALL IN METRIC
OR ENGLISH UNITS, THIS PROGRAM DETERMINES THE KOPPEN
CLIMATE CLASSIFICATION LETTERS FOR A GIVEN GEOGRAPHICAL
LOCATION . . . . KOPPEN

TESTS THE KNOWLEDGE OF FIFTEEN COMMON MAP PROJECTIONS.
THE USER ANSWERS QUESTIONS ABOUT THE TYPE OF DEVELOPED
SURFACE, THE AREA COVERED BY THE PROJECTION, THE
PROJECTION'S PROPERTIES MAPPROJ1

THIS IS A SERIES OF FIVE PROGRAMS WHICH TEACH CLIMATE
CLASSIFICATION THROUGH THE USE OF KOPPEN SYMBOLS. EACH
SUCCEEDING PROGRAM IS PROGRESSIVELY MORE DIFFICULT . . . . . . CLIMAT

$->$

At the end you are sent directly into Command Level with the $->$ prompt. At
this point, the programs are run by typing in their names at the Command Level
prompt.

SUMMARY

The information utilities of The Source and Compuserve are time-share systems.
Many users may be connected at once, but each person sees the computer as if he
or she is the only one using it.

Your identification on a network utility is your account number. You verify
your account number by typing a password. The password must be kept secret.

Once on the utilities the menu structure is used to find your way to the desired
information.

The Do-It-Yourself System
All About Macros

One of the reasons to use a computer, and not a terminal, to access these telecommunication systems, is because a computer with the proper software can save a
great deal of time and effort. The computer can automate and make error-free the
most routine, and often annoying, telecommunication tasks.

Our system's AE Pro software allows you to use complex but easy-to-program
macros which can save hours of keystrokes. Once you have properly set up this
system, features such as complete auto-dial and logon will be available. Just turn
on the computer and the software will take over and deliver you to whatever
PAMS or, in an information utility, whatever database or mode you use the most.

Even if you are not using our system, skim over this section. It should give you
an idea of what macros are all about, and some clues as to what type to introduce
into your own system.

The directory of macros is stored in the "Macro.Lib" file on the AE Pro disk.
Each individual macro-group is stored in a special file. (prefixed on the catalog
directory by an "S" rather than Apple's usual "B," "T," "A," "I," and even
"R" files). Putting a new macro to work on AE Pro is a multi-step process that
should be taken one step at a time. The following directions are meant to supple-
ment the software manual and not to replace it. Read the manual first then consider the rest of this chapter a thumbnail guide.

UPDATE MACRO

As an example, we will show you how to set up a macro system that will audiodial and autolog you onto the Compuserve Information Network.

From the main menu type "M" for Macro and then a "/", which should present a list of titles of the macro groups presently on disk. This list is the contents of the Macro.Lib file.

There should be a macro already present called "Micronet." Choose that one to load. Do this by again typing "M" and then the letter of the macro group, the letter which preceded the Micronet title. After AE Pro displays a message indicating that the macro group is loaded it is time to update it. (If, for some reason, there is no Micronet already defined, load any macro.)

Now type "U" for Update/Review Macro. You will be placed in that mode with the following Macro Command Menu displaying on your screen:

Current loaded:

Phone:

B	=	Baud Rate	300
D	=	Macros display/edit	
E	=	Duplex	FULL
F	=	Data word format	8N1
K	=	Chat mode	OFF
L	=	Load macro from disk	
N	=	XON character	$11 = Q
O	=	XOFF character	$13 = S
P	=	Change macro phone #	
S	=	Set terminal parameters	
T	=	Transpose H/RUB	OFF
U	=	Update from current macro	
W	=	Write macro to disk	
X	=	Exit to main command prompt	
Z	=	Format screen	OFF
$	=	Emulation mode	ON
ʌ	=	Apple CAT port	INT

(If you have an 80-column display the above is shown in a two-column format.)

Now you are going to modify the macro group that is in memory. In order to do that first look at the choices on the above menu and consider which ones you want to change.

The way the choices are presently set up will work on any Compuserve number as that system is presently structured. The only thing you will want to change is the phone number. So, hit "P" and enter the phone number that you call to reach the Compuserve Information Service. But then follow the phone number with the characters "!;" (for example, 555-1212!;). Why that should be so is explained below.

Now choose the "D" for display/edit the macro group. Your screen will now show macro keys 0-9 and ":" and ";" and how they have been defined. If there was a Micronet-titled macro group already on your disk then a lot of the work has been done for you. We will assume, however, that you are working with a blank sheet.

The prompt on the bottom of your screen asks: "Macro to edit? (C = Clear)." Use the "C" option so that you can start with a clean slate.

Now simply hit the "0" key and type in the word "off" and hit return. You have now programmed your 0-macro to be the word "off". So, when you type control-W followed by a 0 on-line, the word "off" will be sent to the host computer.

In such a manner you may program such commands as GO EMA-3 (to pick up electronic mail), GO EMA-4 (to send mail) and so on. Just use your favorite commands. (If you don't have any now, don't worry. By the end of this book you should have plenty.) Don't forget to make one macro your account number and another your password.

Now, program the ";" key. Recall that you placed the characters "!;" following your phone number. These special characters tell AE Pro that when this macro group is selected it should automatically use the ";" macro. We will set up the ";" macro to autolog you onto the CIS utility.

The following macro should work for autolog onto any of Compuserve's own phone numbers. If you use one of the other carriers you will have to modify it to fit. Of course, use your own password.

For macro ";" type:

*** ∧ C%:70000,000'%:PASS.WD'

Type the above as indicated, substituting your own account number and password. Here is what is happening. After dialing the phone number AE Pro is sent to this macro by those "!;" characters, which follow the phone number.

The ******* causes AE Pro to wait a bit. Then the ∧ C tells the system to send a control-C. The "%" causes the macro to wait until the system sends the next character which is a ":". (This is when the system asks "Account number:") After the ":" it will send your account number. The " ' " causes a carriage return to be sent. The "%" waits until the next ":" ("Password:" the system prompts.) Then your password is sent followed by another carriage return.

Now hit carriage return in response to "Macro to edit?" and return to the macro command menu. Use the "W" command to write the macro to disk—call it "mnet."

THE MACRO.LIB FILE

You probably already have the Micronet macros cross-referenced in the macro.lib file. You can tell by hitting "M" from the main menu and then a "/".

But suppose you want to have three various auto-logons to the same utility? Perhaps you would like one macro to call up and send you to EMAIL, and another to just leave you at the main menu prompt. To do this you would have to first define another macro group as detailed above.

Then you would have to save it under a different title—perhaps "mnet2." But then, how do you load that macro from the main menu? To do so you must modify the macro.lib file by adding to it all of the macro titles you will be using.

Use AE Pro's editor to do this. Choose the editor with the "Y" function. Once in the editor use the ".G" function (type ".G macro.lib") to get that file. Then type in a line composed of a letter (unique to that macro group), a slash, what the macro is, a slash, and what the macro's name on disk is. At any time typing a ".H" will get you a full list of the editor's features.

In the case of a macro that, for example, takes you directly into sending electronic mail you might add to macro.lib the line: "G/Micronet-Send Mail/mnet2".

The main thing to remember here is that all macro groups must be written to disk as macro files and the macro.lib file must also be updated to keep track of them all. (The first few times you do this, first make a backup disk!)

COMPLETELY AUTOMATIC

The only thing left to do is to set up AE Pro so that all you have to do is to boot the disk to connect to a system. (Of course, hitting the "D"and then the "M" key is not all that hard to do—but AE Pro makes even this unnecessary.)

Just run the INSTALL program again. This time, choose the "F" or "Macro Action Characters" option. From that menu choose the "Y" for "Default Macro Group At Run" option. Enter the letter of the macro you want to be run automatically. That's all there is to it.

Boot the disk. Sit back and watch as you almost magically log onto the utility. It's a show all in itself.

5. Keeping In Touch– Via Computer

One of the main advantages of being connected to one of the information utilities such as The Source and the Compuserve Information Service is the ability to communicate with other users throughout the country. If you have ever wished that the mail wasn't so slow—or that it wasn't costing a fortune to use more reliable overnight mail services—then the electronic mail services are for you.

On Compuserve the mail is called EMAIL. On The Source it's known as SMAIL. But no matter how they spell it, electronic mail means speedy, reliable delivery of messages and information from you to anyone else on the service.

Services on each utility are similar, though The Source does offer some extra, advanced features. First we'll take a look at the basic features of each service. Then, we'll examine some of the more advanced ways of using both SMAIL and EMAIL.

Compuserve's EMAIL

The easiest way to get to the Electronic Mail area of the Compuserve Information Service is to step through the menu structure. Here is a step-by-step guide beginning at the first, or "top", menu:

Compuserve   Page CIS-1

Compuserve Information Service

1 Home Services
2 Business & Financial
3 Personal Computing
4 Services for Professionals
5 User Information
6 Index

Enter your selection number,
or H for more information.

!1

Note how we choose #1 for "Home Services." This moves us along the menu structure to the "Home Services" menu:

Compuserve   Page HOM-1

HOME SERVICES

1 Newspapers
2 Weather
3 Reference Library
4 Communications
5 Shop/Bank at Home
6 Groups and Clubs
7 Games and Entertainment
8 Education
9 Home Management

Last menu page. Key digit
or M for previous menu.

!4

Choosing "Communications" or choice #4 takes us to the next menu:

Compuserve   Page HOM-30

COMMUNICATIONS

1 Electronic Mail
  (user to user messages)
2 CB Simulation
3 National Bulletin Board
  Public Messages
4 User Directory
5 Talk to Us
6 Lobby Letters of America

Last menu page. Key digit
or M for previous menu.

!1

We will be returning to this menu a few more times in the course of this book to see how some of the other choices make user-to-user communications easy and fun on Compuserve. For now, we will take choice #1 and move to the first screen in the EMAIL area which will display just a short introductory message.

Compuserve   Page HOM-26

Welcome to EMAIL, the user-to-
user message system from

Compuserve. EMAIL allows you to
communicate with other users of
the information service.
Instructions and options are
included on each page. You are
prompted for all required
information. If you are not sure
of what to do, key H (for Help)
and receive further
instructions.

Key S or [Enter] to continue

Hitting return moves us to the main menu of the EMAIL structure.

CompuServe   Page EMA-1

    Electronic Mail Main Menu

1 Read mail
2 Compose and send mail

Last menu page. Key digit
or M for previous menu.

Note how the page number on the screen above is EMA-1. Like any other page
in the Compuserve Information Service this could have been reached directly. If,
at the end prompt of the first menu, "GO EMA-1" had been typed, we would
have been routed directly to the EMAIL main menu. There would then not have
been any connect costs for all of that menu-reading time. This is our first real illustration—there will be many more—of how familiarity with a system will keep
costs down in later accesses.

Keep a little file or notebook of areas in the information utility which you use
most often. By going directly to specified page numbers menus can be bypassed
which, although of excellent utility and quite necessary for the novice user, do eat
up connect time. Remember the cardinal rule of using any information utility:
Time is—literally—money.

Now, let's choose the send mail choice and practice a little bit with EMAIL.
Either via the menu structure, or by typing "GO EMA-4" at the end of any menu,
go to the sending and composing area of EMAIL.

CompuServe   Page EMA-4

Create or edit message using:
    1 FILGE editor
    2 ICS editor
    3 File from disk area
    4 Send message from workspace
    5 Information on FILGE
    6 Information on ICS editor

Last menu page. Key digit
or M for previous menu.

!1

Note that choice #1 has been made. This will let us compose our message using the FILGE editor. The FILGE editor will be more fully explained in chapter eight, "Files On the Information Utilities."

The system responds to our choice with the message:

New file Z99EMA.TMP created — ready

The file-number will be different, but the important thing is to wait until you see that a temporary (.TMP) file is ready before you begin typing. The message is usually very quick, but it is possible, without waiting, to lose the first few characters that you type. And, some days when the system is slower, quite a bit of work could be lost.

Now, just type in any message in any format or design. Here is a sample message:

The Electronic mail system, called EMAIL, on
CIS is easy to use. The FILGE editor allows you
to type text in and
    in
       any
         format.
/ex.

Note the "/ex" at the end of the message on the beginning of a blank line. This is the command which tells the system that this message is composed and ready to be mailed. You will be returned to the EMA-4 page. This time take choice #4 to send the message.

Choosing to send the message will take you to page EMA-8 where the system will ask for the address of the message. We decided to send the message to ourself. This is a good idea for your first message. You will then have an opportunity to check to see that the message comes across the way it was intended and you will have a chance to familiarize yourself with the receive mail functions right away.

CompuServe   Page EMA-8

Key ? for help, or   ENTER
to leave
Send to User ID
:70001,1056
Subject (32 characters max)
:EMAIL on CIS
Your name (32 characters max)
:Neil Shapiro
Is this correct? (Y or N):**Y**

Message awaiting delivery.
Key [Enter] to continue

Obviously, at this point, it would be somewhat of a disaster if only the name of the adressee was known. The system doesn't even want that as secondary information. It is up to you to know the User ID account number (sometimes called a PPN) of the person to whom you want to EMAIL.

Assuming that you have sent a message to yourself, it will take anywhere from a few minutes to a few hours to be delivered. But the next time you call into the system there will likely be a notice that "You have EMAIL waiting" on the screen. Step through the menu structure to get to page EMA-3. Or, just type "GO EMA-3" at one of the end-of-menu prompts.

!GO EMA-3

CompuServe   Page EMA-3

1 Neil Shapiro/EMAIL on CIS

Last menu page. Key digit
or M for previous menu.

!1

The messages that you have waiting for you are listed one-by-one. Each listing contains who the message is from and the message's subject. As above, choose which message you want to read by typing in its number and hitting return.

CompuServe   Page EMA-5

8-Aug-82   11:42   Fr [70003,273]
The Electronic mail system, called EMAIL, on
CIS is easy to use. The FILGE editor allows you
to type text in and
    in
        any
            format.

Key [Enter] to continue

The "header" of the message shows the date and time the message was filed and the account number of who sent it. (This, of course, is the message we just sent.) At the end of the message pressing return will present these choices:

CompuServe   Page EMA-5

1 File this message, then
  delete from mailbox
2 Delete from mailbox
3 Display the message again

Last menu page. Key digit
or M for previous menu.

The last two choices are self-explanatory. The first choice will store the message in a special area called your "files." Once stored there the message may be recalled later, edited, sent to other users. (Again our chapter "Files on the Information Utilities" details all sorts of advanced filing techniques.)

SMAIL ON THE SOURCE

The best way to read or send mail on The Source is to do it from command mode. When you check onto the source go right to Command Level from the choices on the entry screen:

1 OVERVIEW
2 INSTRUCTIONS
3 SOURCE MENU
4 COMMAND LEVEL

ENTER ITEM NUMBER OR HELP **4**

To send a message requires you to type "MAIL SEND" or "MAIL S" at the command mode prompt of " – > ". Once you type this, the system will begin prompting you for input as follows. (Again, a reminder that the convention here is that what the computer types is in regular type face, and what you type is shown in bold type.)

– > **MAIL SEND**

To: **DMO090**
Subject: **The SMAIL System**
Text:
**This test of the SMAIL system shows that the entry of
text is easy**
      ......and easy to format.
**Besides sending simple messages from one user to
another, SMAIL has many other features too.**

DM0090—Sent

Hitting the escape key (or typing ".s" at the beginning of a blank line) will automatically send your letter. Note also that the message is typed in the way it is to be formatted.

Mail delivery on The Source is very fast. So, without even waiting, you can read a practice letter to yourself as soon as the command mode prompt is seen. This is done with the "Mail Read" command.

– > **mail read**

To: DM0090
From: DM0090   Posted: Sun   8-Aug-82   11:18   Sys 13   (5)
Subject: The SMAIL System

—More—

At this point the word "—More—" shows the text is waiting to be read while the header shows who the message was from, the message's subject and the time and date it was posted. Hitting return will display the text of the letter. If, instead, "No" is entered, you will skip ahead to the "Dispositions" choices. "Ne" would move to the next letter, "Del" would completely erase (delete) the letter and "Quit" would return to command mode.

For our practice session let's hit return. The text is displayed:

This test of the SMAIL system shows that the entry of
text is easy
              . . . . . .and easy to format.
Besides sending simple messages from one user to
another, SMAIL has many other features too.

Disposition: **next**

End of mail.

Note that at Disposition we entered "Next" and the system told us there was no more mail waiting for us. But, let's suppose we had many messages waiting. And, further, some would prove interesting and others likely less so. If you were pulling a sheaf of envelopes out of a conventional mailbox you would scan over their return addresses to see who sent the letter within. The SMAIL service also allows you to scan. And, you can not only scan the return address—you can also get an indication of how long is the message's text.

–> **mail scan**

1   From:DM0090   Posted: Sun   8-Aug-82   11:18   Sys 13   (5)
         Subject:The SMAIL System

Read or Scan:

Typing "Read" would let us read the message (which the number in parentheses tells us is 5 lines long), "Scan" would scan the message list again. Another very handy SMAIL command is "MAILCK" which stands for Mail Check. This will tell you exactly what is the state of your electronic mailbox.

– > **Mailck**

MAIL CALL (3 UNREAD, 1 UNREAD EXPRESS, 5 READ: 9 TOTAL)

Note that one of the categories in a Source mail check is "Express." That's just one of the many types of advanced mail techniques on The Source. Compuserve, too, offers some advanced techniques. Here are the ones that are the most useful.

MESSAGE OPTIONS

The Source SMAIL has a special "RE" for Reply command which can be entered at the "Disposition:" prompt. This command saves having to type (or remember) the account number of the person who sent the message to which you are replying.

The SMAIL area of The Source also allows you to Forward a received message to a third party. Simply specify the account number of the person to whom the message is being forwarded.

Many options of SMAIL do not become readily available unless you are willing to enter the abbreviated commands directly from command mode. This is done by typing in a line such as the one below at the command prompt (– >):

– > **MAIL EX TCA234 BC TCA555 CC TCA111 RR**

This command line tells the system how to send the text that will be typed when prompted. The EX stands for sending it Express to TCA234 (he will see it listed as Express during a Mail Check); sending it as a Blind Carbon to TCA555 (other people will not see that he got a copy too); the CC sends a normal carbon-style copy to TCA111 (listing him as a recipient of a copy); and the RR simply means that replies are requested. You may use all or any of these SMAIL commands in any combination.

You may also, on SMAIL, use the on-line text editor (see "Files On The Information Utilities") to create a special file called MAIL.REF. In this file can be a list of people's account numbers and their corresponding names. Using this program as an automatic addresser you can send form letters to a group of people. Up to 200 account numbers can be included in one list and any number of lists may be stored in your file space (subject to the current storage costs).

EMAIL, on Compuserve, offers a less automated ability to continue sending a letter—once it is composed in your workspace—time after time. All that must be done is to keep choosing the "Send From Your Workspace" option and, each time, enter a new account number.

Both Compuserve's EMAIL and The Source's SMAIL allow composing letters outside of the actual mailing program and storing them in your account area. Then, once in the mail area, you can send one of these pre-stored letters. The main advantage to this is that such letters can be composed while off-line using a word processor or the editor built into your terminal program. Utility time and costs can be saved in this manner.

Once the letter has been word-processed to your own satisfaction, it should be sent to the utility and stored. This process is called uploading. The uploading process is generally demonstrated in the next chapters, and the how-to section will focus on a step-by-step method using our example system.

And, both The SMAIL and EMAIL areas allow storing incoming mail in your file space so that it can be later edited, and even downloaded. As uploading means to send information from your computer to another (in this case the utility's mainframe computers), downloading is the opposite. Information downloaded comes from the utility to your computer where it can be stored on your own floppy disks without storage costs, and where any such file can be used at leisure. This process too is detailed in the following chapters.

6. Communities of Information

In the last chapter we detailed enough information on the electronic mail services of the two major information utilities to enable you to begin using those services. But, are there ways to reach the entire community of system users, without this laborious and time-consuming process of typing each account number? Can you reach the huge electronic mail audience with a message quickly and easily?

The entire electronic mail audience can in fact be reached through use of the community-information areas on either of the utilities. On Compuserve either the most appropriate Special Interest Group or the National Bulletin Board could be used. On The Source you might turn to their Participate area and join—or begin—a conversation devoted to the topic; or the Post could be used for a general note.

These areas—Compuserve's Special Interest Groups (called SIGs) and the National Bulletin-Board; and the Source's Post and Participate—deserve our special attention. They are the means for contacting literally hundreds of thousands of system users directly. And accessory programs such as the CB Simulator on Compuserve and CHAT on The Source enable you to contact people on real-time and almost face-to-face basis. These programs are a gateway into the world's first community that has no geographic borders.

Compuserve's Community

THE SIGs

The Special Interest Groups or SIGs are one of the fastest-growing and most exciting areas on the Compuserve Information Service. Over thirty groups cover interests which range from various computers to different computer languages; from space exploration to experimentation with the very concept of newspapers and magazines.

All of these areas share the same operating system, so it is only necessary to be familiar with one to know your way around all of them. Many people on Compuserve find that they spend the majority of their on-line time in these popular groups.

The SIGs are found in two main areas on Compuserve—the Home Information area and the Personal Computing Area. The SIGs can be reached by menu or command.

By menu you would first choose either Home Services or Personal Computing from the main, or top, menu. From there choose the "groups and clubs" selection. This would take you to either of the following two pages:

CompuServe   Page HOM-50

GROUPS AND CLUBS

1 CBIG
2 Hamnet
3 Netwits
4 Photo-80
5 Sports
6 Cooks' Underground
7 Belmont's Golf SIG
8 Instructions
9 Descriptions of Groups

CompuServe   Page PCS-50

COMPUTER GROUPS AND CLUBS

1 CP/M Group	8 MNET80	TRS-80
2 HUG (Heath)	9 LDOS	TRS-80
3 MAUG (Apple)	10 VTOS	ST80
4 MNET-11(H11)	11 MCONN	TRS-80
5 MUSUS = PASCAL	12 CLUBIG	
6 RCA Group	13 AUTHOR'S SIG	
7 TRS-80 Color	14 Commodore	

17 Instructions   18 Descriptions

All of these SIGs are on-going, message-based conversations. The main idea is to first scan through the few hundred messages that are always on a SIG. Then read the ones most likely to appeal to you. Replying to particular messages allows you to join ongoing, electronic correspondence. A message and all of its replies is called a "thread." Some threads become dozens of messages long as many users add their own subsidiary messages. Other threads are short, and may contain only a few messages. Of course, anyone may leave an original message and begin one of these threads.

All the messages on a SIG will have something to do with that SIG's main topic. The messages on HUG (Heath Users Group) will all be somehow related to the Heath line of computers. Messages on the Cooks' Underground will all have something to do with recipes or the art of eating.

At this point choose a group or club which has some appeal for you. We will examine the MAUG or MicroNET Apple Users Group here, a group of special interest to users of the Apple computer. But you may use our explanations of commands and strategies in any of the other SIGs. The fact that these message boards all use the same software is one of the strongest features of the SIGs area—learning one means you have learned them all.

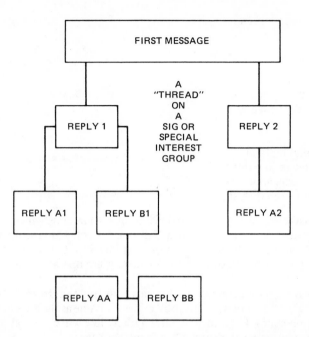

A "thread" on one of Compuserve's Special Interest Groups (SIGs) is shown. The first message is replied to by two other users, the second reply also picks up another message. Then, one of the secondary replies receives two further messages. In this way, a thread of conversation becomes a long, and interesting topic.

After we choose a SIG from the menu, we get the usual message to wait. This few-second wait transfers us to that host computer which is running the SIG software. The screen then shows:

Thank You for Waiting
Your name: **Computerist**

Welcome to MAUG, V. 1A(43)

System contains messages
13584 to   14253

Brief bulletin:

If you are not yet a MAUG member, please note that only members may leave messages other than a membership request addressed to SYSOP. Once your request has been processed you will be able to leave messages to the group. Type MI at "Function" for more information.
Best,
Neil Shapiro (Sysop) 70001,1056

Enter blank line for menu:

MAUG
Function menu:
1   (L)   Leave a message
2   (R)   Read messages

5   (B)   Read bulletins
8   (MI)   How to join this SIG
9   (OP)   Change your SIG options
0   (E)   Exit from this SIG

   As shown above, the first thing to do is to enter your name which "introduces" you to the system. Some of the SIGs will then ask if you want membership and, if you do, will automatically add you to the membership. Others, such as MAUG, require that you leave a message to the Sysop (system operator) asking to become a member.
   Leaving such a message requires choosing that function from the short menu. Answer the "To" prompt with "Sysop" and your name as the answer to "From." Request membership. Hit return to enter a blank line when done and then, at the "option" prompt enter the letter S for Send Message.
   Once your membership has been validated on a SIG, the next time you check in you will see a screen similar to:

Welcome to MAUG, V. 1A(43)

CompuServe sponsor: BILL LOUDEN 70006,111

Name: Your Name, Your Account No.
Last on: 03-Aug-:0   20:00:00
High msg-: 0

You are user number 38651
System contains messages
13584 to 14253

Brief bulletin:

***Welcome to the SIG!

If this is your first time on, please enter   MI   at the
"Command:" prompt to receive Membership Information. Read the
bulletins with the [B] command.

***The special MAUG CB may be used at any time by entering the
[CO] command. We have established 10pm EDT, however, as a standard
"calling" time.

***MAUG weekly meetings are now held on this SIG's CB every
Sunday at 8pm EST on channel 30. All members are welcome!
Happy Apple'ing!
Neil Shapiro
MAUG Sysop 70001,1056

Enter blank line for menu:

MAUG
Function menu:
1   (L)   Leave a message
2   (R)   Read messages
3   (RN)   Read new messages

5   (B)   Read bulletins
6   (CO)   Online conference

9   (OP)   Change your SIG options
0   (E)   Exit from this SIG

    The first thing to do is to change your options to command mode from menu mode. The SIGs are one of the few areas on the information utility where the menu structure is not nearly as powerful—or as easy to use—as straight command mode. When you choose this option you will be in the:

User option menu:
1   Change to command mode
2   (LL)   Change line length
3   (T)   Return to Function menu

0   (P)   Make options permanent

Enter selection or H for help: **1**

    Entering the choice 1 instantly changes the screen to command mode and you see:

User option:

    Hitting return produces a list of these options.

T   —   return to Function level
ST   —   stop between messages (*)
NS   —   don't stop between messages

LL   —   change line length
BR   —   set brief mode, which suppresses
            repetitious display of options
NB   —   clear brief mode (*)
DS   —   set default login Section (O)
P   —   make options permanent

MEN   —   use menus instead of command prompts

User option:

The one other thing to do right away is to set the line length if you are using an Apple II with an 80-column card. No matter what you choose, don't forget to also use the P option to make all this permanent so it all won't have to be done again. Finally, a T sends you the the "Function" level of the SIG. It is from this Function level that most of your commands will be given.

Use H for list of valid commands.

Function: **H**

Functions:

D	—delete	E	—exit
G x	—go pg. x	E(x)	—exit&do x
I	—instructions	L	—leave msg.
NEW	—new/chngs.	OP	—user opt.
QS	—Quick scan	R	—retrieve
R x	—run SIG x	RT	—Read Thread
S	—scan hdrs.	SD	—Scan&Dsp.
SEN	—Send msg.	SN	—Sec. Names
UST	—Cur. users	V	—interests
X	—database	XA	—X ACCESS

? x   —explanation of function x

There are a number of options available. The SIGs are complex structures and it will take some practice before being able to use all of their special features. For now, we will just be concerned with a few of the most basic commands. Using the I for Information command will display a fairly detailed set of guidelines. You can order a SIG manual from Compuserve which details all of the many in and outs— or just use the SIG areas and depend on fellow users for help.

You will also want to use the B for Bulletin option. The Bulletin, usually written by the SIG's system operator (Sysop) more or less sets the tone of the SIG. Here the Sysop usually details just what type of messages will likely be found on the SIG, and any of the SIG's special or unique features that might otherwise be missed.

As a SIG is a collection of messages, the first thing you will want to do is to get an idea of how the messages are arranged on the SIG. The best way to do this is with the scan commands.

Entering S for Scan will list out the messages by their complete "headers."

Function: **S**
[F]orward, [R]everse, [A]bort: **F**
System contains messages
13619 to 14291
Starting message number
(N implies since last time on): **14000**

```
#:    14000         Sec. 2 — Software (General)
Sb:   PRO problem
      30-Aug-82   23:21:09
Fm:   Robert Seaver 70405,1525
To:   Jonathan Teller 70275,1235 (X)


#     14001         Sec. 0 — General
Sb:   #Cat/des
      30-Aug-82   23:38:51
Fm:   Jonathan Teller 70275,1235
To:   Neil Shapiro 70003,273 (X)


#:    14003         Sec. 0 — General
Sb:   AIRPORTS & DISKS
      31-Aug-82 00:16:23
Fm:   Allan Turoff 70315,1221
To:   Adnan Ahmad 72415,1420 (X)


#     14005         Sec. O — General
Sb:   ITOH Graphics
      31-Aug-82   00:24:34
Fm:   Allan Turoff 70315,1221
To:   James Okubo 70265,1304 (X)


#:    14007         Sec. 3 — Hardware
Sb:   #13961-Help needed
      31-Aug-82   00:48:14
Fm:   Jim Tryon 72065,314
To:   Bill Steinberg 70215,1126 (X) ∧ P
```

The complete header carries the message number, the section of the SIG the message is in (if the SIG has sections), the subject of the message (if the message is a reply, the subject is preceded by the number of the message being replied to), the date and time the message was left, who left the message, and who the message is addressed to.

As there are hundreds of messages on the SIG, it would take a long time to scan through them all in this way. An even more abbreviated scan can be done with the QS or Quick Scan method. Use of this command is shown below:

```
Function: QS
System contains messages
13619 to    14291
Starting message (N for new): 14000

14000:  PRO problem                Sec. 2 — Software (General)

140001: Cat/des                    Sec. 0 — General
        1 reply
```

14003: AIRPORTS & DISKS   Sec. 0 — General

14005: ITOH Graphics   Sec. 0 — General

14007: Help needed   Sec. 3 — Hardware

14008: TEX

This just gives the message number and its title and section. It also indicates the number of replies in the thread.

This and the preceding scan were interrupted by entering a control-P from the keyboard. A control-P will almost always return you immediately to the main function prompt.

One other scan-oriented command is the SM or Scan-and-Mark command. This command works like the S command but pauses at the end of each message header. The message can then be marked for reading in full later, or you can skip to the next header. You can also go back to the top function mode:

Function: **SM**
[F]orward, [R]everse, [A]bort: **F**
System contains messages
13619 to 14291
Starting message number
(N implies since last time on): 14000

```
    #: 14000   Sec. 2 — Software (General)
   Sb: PRO problem
        30-Aug-82   23:21:09
   Fm: Robert Seaver 70405,1525
   To: Jonathan Teller 70275,1235 (X)
```

[M]ark [T]op:

```
    # 14001            Sec. 0 — General
   Sb: #Cat/des
        30-Aug-82   23:38:51
   Fm: Jonathan Teller 70275,1235
   To: Neil Shapiro 70003,273 (X)
```

[M]ark [T]op: **5**

If you have marked messages using SM, then you can use the RM (Retrieve Marked) command to read the text of the messages so marked. Otherwise use one of the R for Retrieve commands. Forward or Reverse from a message number can be specified as can an Individual message (RF, RR or RI). To be really fancy, use the RT command and retrieve messages by "threads." Rather than retrieving in numerical order, RT retrieves in a logical order of original message and that message's replies.

When you first check into a SIG, plan on spending a few hours there to catch up on all of the preceding messages. In the future, though, things will be much faster.

You will be able to use the RN command—the Retrieve command with the New option. This will automatically retrieve only the messages new to the SIG since you were last on it. The New option can also be used with the Scan command (SN).

Whenever a message is retrieved you will be presented with a choice of options. The T option will allow you to return to the top function level of the SIG, hitting return will retrieve the next message in line, D will delete the message (if it is one that you originally sent), and using the RE option will allow you to reply to the message being read.

Both the L for Leave and the RE for Reply commands use the same editing functions to help compose your message. The difference is that the L command is for leaving an original message on a new subject. The RE command simply replies to an already existing message and keeps the same subject in the header.

Unlike the EMAIL and FILGE editors (detailed in the next chapter) messages you leave are line-numbered as they are left. These line numbers are used to edit the message before it is sent. Here is a quick look at using the L function to leave a message on a SIG (the RE would be the same but without having to specify any information other than the body of the message.)

Function: **L**
To: **All**
Subject: **Hello!**
1:
Just joined this SIG and I am particularly
2:
interested in telecommunications with the Apple. I have not
3:
used many of the terminal programs. Which do you recommend?
4:

Leave options: **S**

Section #required
0 — General
1 — Software (Games)
2 — Software (General?
3 — Hardware
4 — Programming
5 — Community Square
6 — For Sale/Wanted

Enter selection: **0**

Message # 14299 Stored

The process of composing a message is quite straightforward. The S option, followed by the choice of SIG section, stores the message on the SIG so that all may read it.

Other editing options besides S include the E for Edit, which option puts you into a simple line-editor so that the wording of a message may be changed. The P for Preview option will type the message out on the screen so that you can be sure of it before storage. The I for Insert option allows the insertion of lines into the message. The C for Continue option allows you to tack on additional lines to the end of the message. If you change your mind completely about leaving the message the A for Abort option will return you to the SIG's top function prompt.

Using the SIGs can be both a rewarding and exciting experience. The power of telecommunications will become apparent as the boundaries of contact with other computerists become the interests that you share rather than the geographical distances between you.

NATIONAL BULLETIN-BOARD

This area of Compuserve, the Bulletin-Board, can be reached by using the command GO-HOM30 at any menu prompt (or the command R BULLET from the command area). Once there you will be in a very free-form area of public messages. These messages are divided into three broad kinds of categories: Items For Sale, Items Wanted To Buy, and Posted Notices. The categories of Sale, Wanted, and Notices are searched by entering keywords. Searching for a used TRS-80 computer, for example, might be done by entering the command SCAN SALE TRS-80. The above would provide the numbers of all messages on the board that offered for sale anything with TRS-80 in the message title.

The commands are quickly learned via the board's built-in help function. Just enter "help" at the asterisk prompt.

*help

```
The CompuServe Information Service National
Bulletin Board is an information exchange
medium.
The following commands are implemented:
SCAN      READ        INDEX    VIEW
          COMPOSE     POST     EDIT      ERASE
          CHECK       EXIT     AGE       OFF
Commands may be abbreviated to the first 3 letters.
For additional information about a command, type:
HELP command
For example:    HELP SCAN
```

Here is, in a very brief form, how to use these commands while on the National Bulletin-Board. The Board's own Help commands will further describe them if you get stuck.

Scan requires entry of the kind of category (Sale, Wanted or Notice) and an optional keyword. The format is SCAN KIND KEYWORD.

Read lets you read the messages that you have found with scan. It is done in the manner of READ # — the number being the message number.

Compose allows you to write your own messages. It puts you in the FILGE editor (see last chapter on EMAIL as well as our next chapter on filing). After the message, the last line must be /EX to return to the Board's command prompt.

Post allows you to put your composed message on the board for all to see. The format for the command is POST KIND KEYWORD. If you have composed a message asking for help in an Apple adventure game it might be sent with POST NOTICE APPLE.GAME. Notice how the two words "Apple" and "Game" are run together with a period. This is a convention enabling you to use two words or more instead of one.

Edit allows you to edit any message in your "workspace." That is, you may edit any message just composed or read.

Erase will erase a message that you have left. The format is ERASE #.

Index lets you see the keywords available in a particular category. Using IN-DEX NOTICE would produce a sorted list of the keywords you could search the NOTICE category on.

View lets you read all messages in a category: for example, VIEW SALE.

Check lets you see the status of any of your messages. You can see how many people have read it. Format is CHECK KIND KEYWORD and the keyword is optional.

Age is a helpful command. AGE 2, for example, would set up so that all the commands only referred to messages two days, or less, in age.

Exit gets you off the Board and back to the menu or command mode (whichever you arrived from).

Off signs you off the system.

Each of these commands are slightly more detailed than presented here. But this is enough to get you started using the National BB. You'll find it to be a regular potpourri of information about almost everything.

CB SIMULATOR

Compuserve offers a unique and always interesting way to get to know your fellow computerists. Their CB Simulator allows for conversations, in real-time, between any number of users.

As the name implies, this simulator sets up channels for discussion. You can access the Simulator easiest from command mode just by typing R CB. Or, you can get there from the Communications choice on the Main Menu.

Here's what you see when you arrive:

CB Simulator Ver 3(34)
What's your handle? **Redeye**

Most people go by "handles" on this simulator. Some of the handles are quite colorful. And many people are known to a wider circle of friends by their handle via Compuserve than they are known by their name in "real" life. (As shown, my own handle is Redeye—which refers to how I feel after my early morning commuter train ride from Long Island to Manhattan).

The first thing to do is to type /HELP. Here's what you'd get to read:

```
/help
Prefix commands with /
/TUN 16 — Tunes channel 1 to 36
/MON 1,4 — Listen to extra chnls
/UNM 7,3 — Unmonitor channels
/STA — Type channel status
/EXIT — End session
/WHO — Type PPN of last talker
/HAN — Change handle
/SCR xyz — Scramble on key "xyz"
/SMC xyz — Scr & Monitor Clear
/XCL xyz — Xmt CLear; unscr rcvr
/UNS — Unscramble (both clear)
/SQU abc — Squelch handle "abc"
/JOB — Type your job #
/HELP — Type this message
/TALK — Go to two person talk
/SPCWAR — Go to space-war
/USTSAT — Type user status
```

The commands all relate to the channelized structure of the CB. The /STA command reveals the status of the channels. Typing /STA might reveal something like:

(2)5# (12)6 (19)2* (31)4

The number in parentheses are the channel numbers. Five people are talking together on channel 2, six are talking on channel 12, etc. The # symbol shows that you are "tuned" to channel 2. This is your main channel. You are receiving the conversation and this is the channel you will "transmit" on when you type. The * shows that channel 19 is being monitored. You are reading the conversation there, but could only answer by using the /TUN command to reset to that channel.

How easy is it to begin? Very easy indeed. One afternoon we just got into the CB and /TUNed to channel 3. We were alone but in a very few minutes:

(3,**CPA**) hi

The person calling us had the handle of **CPA** and the 3 showed that he was also /TUNing channel 3. We answered:

Hi CPA

and another CB conversation was initiated:

(3,**CPA**) how ya doin
Good . . . Hey, you want to be in a book? I'm capturing this for a book I am writing . . .
(3,**CPA**) sure why not
Great! Where are you? We're in New York.
(3,**CPA**) wash dc here
What's your computer? (Apple here)
(3,**CPA**) trs 80 mod 3
(3, SANDINISTA!) HELLO ALL!

At this point, the conversation suddenly became a three-way...

Hi Sand . . . . . wanna be in a book? . . . I'm capturing this.
(3,**CPA**) hi sandi
(3, SANDINISTA!) RED   WHATS THE CLOCKSPEED ON AN APPLE?
Sand=> 1MHz
(3, SANDINISTA!) SURE. CAPTURING W/BUFFER?
Sand=> Yep...
/sta
(3)3 #
(3,SANDINISTA!) THANKS, I'D LOVE TO BE A HOUSEHOLD WORD.
Sand = > Where are you located?
(3,SANDINISTA!) IN LOUISVILLE, KENTUCKY
OK . . . I gotta run. Book is called Small Computer Connection.
Maybe you'll see yourselves!
(3,SANDINISTA!) CPA   WHERE ARE YOU?
(3,**CPA**) wash dc
(3,SANDINISTA!) THANKS
(3,**CPA**) bye Red
/exit

The CB Simulator is one of the most popular areas on the entire information service, according to Compuserve. One day you can find yourself talking to a fourteen year-old student, a housewife, a nuclear chemist, a famous author, a sports figure—the list is endless. It is one of the most interesting ways to meet and interact with others—a futuristic way to make lasting friendships.

Community on The Source

THE SOURCE'S PARTICIPATE

Similar in some ways to the concept of the Special Interest Groups (SIG) on Compuserve; the Source's Participate area does some things better and some things not as well. You may find that Participate will be quite helpful. Some users, however, find Participate somewhat lacking in user-friendliness.

The idea behind Participate is an exciting one. It is a free-form computer conferencing area. A conference may be started by anyone, and made either public or private. Public conferences may be joined by all; private conferences are joined only by invitation.

As on the SIGs of Compuserve, users send and reply to messages all of which are stored on the utility's disks. In this manner, long and involved conversations can be carried on.

One important thing to note is that—in Participate—the person who starts the conference is responsible for paying the storage (at the Source's current rates) on disk of all that conference's related messages. But, there are many conferences in which this is not a factor—conferences that were begun by the Source, for example. If you are planning on using the Participate area by beginning conferences on your own, be aware of this cost structure.

The Participate area is reached by following the "Communications" path on the Source's Main Menu (or by simply entering the command PARTI from the main command area):

THE SOURCE MAIN MENU

1 NEWS AND REFERENCE RESOURCES
2 BUSINESS/FINANCIAL MARKETS
3 CATALOGUE SHOPPING
4 HOME AND LEISURE
5 EDUCATION AND CAREER
6 MAIL AND COMMUNICATIONS
7 CREATING AND COMMUNICATIONS
8 SOURCE*PLUS

Enter item number or HELP **6**

MAIL AND COMMUNICATIONS

1 MAIL
2 CHAT
3 POST
4 PARTICIPATE
5 MAILGRAM MESSAGES

Enter item number on HELP 4
Welcome to PARTICIPATE on The Source

PARTICIPATE ON THE SOURCE
----
1 Overview
2 Instructions
3 List principal conferences
4 Begin to PARTICIPATE

Choose one (or QUIT):

The best bet at this point is to take choice 2 and carefully study the long and involved instructions. Here is a little bit of what is needed — enough to get you started. But, more than any other database structure we know of, Participate will require spending much time learning and using the commands before you can obtain any degree of proficiency.

The very first thing you will do when using Participate is to answer a one-time-only questionnaire. This will customize Participate's output to your own computer's line length, and other such variables.

Once the questionnaire is completed, you will be introduced to your "mailbox." The mailbox is where you will be receiving invitations to join either public, or private conferences. The first time on, there will be two messages awaiting you:

2 NOTES WAITING
READ, SCAN, BATCH, CANCEL OR HOLD?

On typing the command "Read," you will receive your first invitation—to join PARTI. The invitation will end (as all such will) with: "Join to receive future answers?" You had better answer "yes" to PARTI because all the other conferences are part of this main one. Then you will be told that you have joined and the second invitation—to a conference called PRACTICE—will be offered. It would be a very good idea to join the PRACTICE conference, a tutorial which will take from a half-hour to a few hours to complete. But, for now, answer N (for Next) which will take you to the main prompt of the Participate program.

READ, SCAN, MODIFY, WRITE, JOIN, LEAVE, ORGANIZE, READ ABOUT, HELP, QUIT?

ACTION?

The Action Prompt will allow you to begin finding your way around the Participate section. The first thing is to answer the prompt with READ ABOUT 1 CONTENTS (or R A 1 C). This will display the names and locations of all the other conferences that go to make up Participate.

A long list of lines will be displayed: one of which will look like:

1 "Facilitators", 82.2 (56)

The number at the beginning of the line tells the position of the note which began the sub-conference within the Participate area. Number 1, in this example, means it is the first sub-conference—and the word "Facilitators" is the sub-conference's title.

Following the title of the sub-conference is the sequential number of the note which began that subconference. The number is composed of the year the note was left, followed by a decimal point, followed by the number of note it was within that year. The 82.2 means that this was the second note left in the year of 1982.

The number in parentheses is the number of answers that are already within this sub-conference (the number will, of course, change with almost every logon).

When the list of public conferences has scrolled by (or with control-P or break) you will be returned to the Action prompt.

Now, to read one of the conferences, use the R command. Let's say that one of the conferences was:

656 "Free Software," 82.4532 (128)

and that you were interested in this free software. At the Action prompt you could then type R 82.4532 which would display the note which begins the conference. Then, after reading the first note in the conference, a new prompt would appear saying:

Disposition of "Free Software?"

You could then type READ 1-X (an example might be READ 1-20), where X is the number to stop at. The command READ 1-## (entered just as shown) is a handy one to remember because this is the command to read every message in a conference.

As you can see, there are two main prompts in Participate—one is the Action prompt and the other is the Disposition prompt. The Disposition prompt is received whenever a message is finished. From the Disposition prompt there are a number of options.

The R for read command can be used in a number of ways. Just entering R by itself will reread the note. Or, entering R 4 would, for example, read the 4th note in the conference. Entering R 4-18 would read notes four through eighteen.

There is also a R A command (Read About). This can have four different endings attached: R A J stands for Read About Joiners and tells the names of others in the conference. R A A is used in private conferences and describes the people to whom the conference is addressed (Read About Addressees). R A C stands for Read About Conference and that command will give you a summary of the conference's goals or reason for being. Finally, R A A for Read About Author, will reveal who began the conference.

If the note you are reading is the one which began (established) the conference, then entering J for Join will display the subsequent responses.

To go to the next note, simply type N for Next.

To reply to the note, type W for Write. With this command you go to what is called the "scratchpad" area." The scratchpad is a temporary filespace in which the note can be composed. The scratchpad works the same way as the workspace in SMAIL (see last chapter). Type in the message just as you want it sent.

When through composing the message you will want to send it. There are three ways to do this. All are done by typing in the options as the last line of the message.

Typing, as the message's last line, .CONFER (don't forget the period preceding) would set up your message as the beginning of a new conference. If you do this you may follow the command with the participant's names (if private); or the name of the existing conference in which case your conference becomes a sub-conference.

Ordinarily, you will use the .ANSWER option which will post your message as a typical part of the conference. There everyone who joins the conference will be able to read it and—if they want to—respond.

Or, you can send it with the .MESSAGE option. In this case only the person to whose note you are replying can read your response.

Of the three options, in the ordinary conference, the .ANSWER option will likely be used the most.

At this point, it is necessary for us to give one warning about Participate which—at the time you read this book—may or may not still be true. We found the response times on Participate to be very slow. It was not unusual to request a message and then wait up to three minutes for it. Considering that the user pays for this connect time, the problem might mitigate against heavy use of this area. However, we have been told by Source spokespeople that the response will be improved. By this time, if you have been following along at your own keyboard, you can judge if the response times are now suitably fast.

Participate is not an easy area to learn. But once learned, you may find it to be one of the system's most useful areas. If you're a hobbyist—here are the interactive users' groups. If you're a businessman you can set up your own private conferences only for the people working for you and relating solely to your own business interests.

Remember — Participate requires a certain amount of practice on your part before it will become second-nature. At the prompts you can always enter a question-mark (?) to receive some help as to what to do. But the best thing is to spend some time — at least an hour — in the PRACTICE conference.

THE POST

The Post is The Source's Bulletin-Board system. It is a very powerful program which makes full use of both categorization and searching functions so that leaving and reading notices is easy. It contains more advanced features than the Bulletin program on Compuserve.

The Post is reached via the communications choice of the Source's Main Menu:

THE SOURCE MAIN MENU

1 NEWS AND REFERENCE RESOURCES
2 BUSINESS/FINANCIAL MARKETS
3 CATALOGUE SHOPPING
4 HOME AND LEISURE
5 EDUCATION AND CAREER
6 MAIL AND COMMUNICATIONS
7 CREATING AND COMPUTING
8 SOURCE PLUS

Enter item number or HELP **6**

MAIL AND COMMUNICATIONS

1 MAIL
2 CHAT
3 POST
4 PARTICIPATE
5 MAILGRAM MESSAGES

Enter item number of HELP **3**

POST

1 OVERVIEW
2 INSTRUCTIONS
3 READ
4 SEND
5 PURGE

Enter item number or HELP

The three choices as to activity — Read, Send and Purge (Delete) — all use basically the same command structure. And, they are all equally dependent on the categorized structure of the Post and on its searching facilities.

The Post is divided into categories. Within each of the categories, messages are further tagged with a keyword. This allows for a great degree of selectivity.

For example, let's choose to Read the Post.

[R]ead, [PO]st, [PU]urge, [S]CAN, OR [H]elp: **R**

[C]ategory, [U]ser ID, [D]ate, [K]eyword:

At this point, choose the major division of information as to which notices you want to read. There are more than 70 categories (listed below) to choose from. You can read notices posted by a specific user. You may choose to read notices posted during a specified time limit. Or, you can use a keyword which, in effect, looks at the sub-categories of the main category choices.

A list of categories is available at the Category prompt:

[C]ategory, [U]ser ID, [D]ate, [K]eyword: **C**

Categories, or [H]elp: **H**
AIRCRAFT
ANTIQUES
APARTMENTS-RENT
APPLE
ART
ASTROLOGY
ATARI
AUTOMOBILES-DOMESTIC
AUTOMOBILES-FOREIGN
AVIATION
BASIC
BULLETIN-BOARD
BUSINESSES
CADO
CBM/PET-COMPUTERS
CHATTER
CLUBS
COLLECTABLES
COMMODORE
CP/M
CPT
DATING
DEC
DOCUMENTATION
ENGINEERING
FAIRS-AND-FESTIVALS
FORTRAN
GAMES
GRIPES
HAM-RADIO
HARDWARE-RENT
HARDWARE-SALE

HAYES
HEATH
HELP-WANTED
HEWLETT-PACKARD
HOBBIES-AND-CRAFTS
HUG
IBM
INFOX
LANIER
LIBRARY-FORUM
MERCHANDISE
MUSIC
NEC
NOVATION
OFFICE-EQUIPMENT
OSBORNE
OSI
OVERSEAS
PARTI
PASCAL
PERSONAL
PETS
PHILIPS
PHOTOGRAPHY
POLITICS
PROPERTY/HOUSES-RENT
PROPERTY/HOUSES-SALE
PROPERTY/HOUSES-SWAP
PROPERTY/HOUSES-WANTED
PUBLIC-FILES
PUZZLES
RKO-FORUM
SATELLITE-TV
SAYINGS
SERVICES
SOAP-OPERAS
SOFTWARE-SALE
SOFTWARE-WANTED
SOURCE
SPORTS
STEREO/TV
STUDENTS-CORNER
TI-99/4
TRAVEL
TRS-80
USER-PUBLISHING
VIDEO
VISICALC
WANG

WEEKEND-GETAWAY
XEROX
ZENITH

Reading a notice first requires choosing a large category and then zeroing in on the item of interest:

POST

1 OVERVIEW
2 INSTRUCTIONS
3 READ
4 SEND
5 PURGE

Enter item number or HELP:**3**

[R]ead, [PO]st, [PU]rge, [S]can, or [H]elp:**R**

[C]ategory, [U]ser ID, [D]ate, [K]eyword:**C**

Categories, or [H]elp:SOFTWARE
Searching . . .
24 notices valid.

[N]arrow, [E]xpand or Return for all: **N**

[C]ategory, [U]ser ID, [D]ate, [K]eyword:**K**

Keyword(s) or [H]elp:APPLE
Searching . . .
3 notices valid.

In the above example, we narrowed our search down to only three valid notices (We went from notices tagged Software to those tagged Apple Software). The more familiar you get with the Post, the more exacting your searches will become.

The Scan and Send functions are very much the same as the Read function as far as prompts and answers. The only thing left to try is the expert mode. This is done from Source Command Level (with the −> prompt, not the menu structure). You could read notices by simply combining all commands on one line. For instance: POST READ SOFTWARE APPLE would bypass all the menus. Once familiar with the system this method is a great time saver.

CHAT ON THE SOURCE

The Source allows two users on the same system to talk real-time to each other via their keyboards. Unlike the CB Simulator on Compuserve which allows for (indeed encourages) free-wheeling conversation among dozens of users — the Chat function is limited to only two users.

To chat with someone on The Source requires knowing their account number. You can find fellow chatters in many ways. One way is to use the DISEARCH command (available from the Command Mode) which provides a list of subscri-

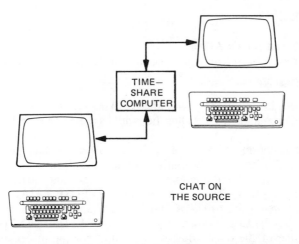

CHAT ON
THE SOURCE

Chat allows two people on The Source to type back and forth to each other. Either can initiate or end the conversation using simple commands.

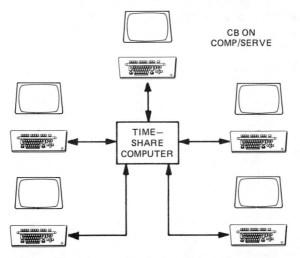

CB ON
COMP/SERVE

CB on Compuserve allows dozens of users to type to each other. Conversations are completely open-ended and a person may be in more than one conference at a time.

bers to The Source and what they have listed as their interests or hobbies. Or, use the CHAT category of The Post (detailed above). Typing ONLINE brings a list of other people (by account numbers) who happen to be on at the same time as you. (This last is a little tricky as you may wind up calling someone who either does not know how to CHAT, does not want to be bothered, or both.)

Let's say that you know your friend, account number STX666, is supposed to be on. You type ONLINE (all of this while in Command Mode), and see that his account is one of the ones listed.

Then, just type something similar to:

CHAT STX666

Hi Old Buddy, can you CHAT?

He would see on his screen your account number in the following format:

DM0090 USER 20

If he had been in a program he would have to use the control-P or break key to get back to Command Mode. Then he would type CHAT and your account number.

Chatting then continues back and forth until either user hits the escape key or types .S and RETURN at the beginning of a line (don't forget the period).

By the way, if you will be doing something very important and do not want to be interrupted by random CHAT requests, type REFUSE CHAT at Command Level. You can even type REFUSE CHAT PERM. Both options can be reset from either type of no-chatting option by the command REFUSE NO CHAT. CHAT-ting can be one of the most addictive of things for the information-oriented extrovert.

No matter if you prefer the SIGs on Compuserve or PARTI on The Source, if you like CB or CHAT—you'll find a whole new world of personal acquaintances opening to you. Telecommunications means not only computer-to-computer information exchange; but person-to-person contacts too.

7. Surveying the Utilities

Telecommunications has two modes. The first mode, the one with which this book is primarily concerned, involves using the art of telecommunications in an interactive manner. The information utilities, as we have seen, provide for swift and sure communications. And, as we will investigate in the next chapters there are even more astonishing interactive goals which may be achieved by setting up privately-run, small networks.

The other mode of telecommunications is a more passive one. The utilities we have already looked at—The Source and Compuserve—offer hundreds of databases online which cover virtually every consumer field of interest. Other utilities such as Dialog and the Dow Jones News/Retrieval Service offer their own style of databases oriented more toward the businessperson and the professional.

In all of the thousands of databases available, the odds are that everyone can find whatever information that might be required. Searching through the databases for information is the second way that telecommunications can be utilized.

The Source and Compuserve deal with an eclectic assortment of information providers. These information providers make their data available to the Utilities who in turn make it accessible to their online customers. The information providers receive a portion of each customer's connect-time spent in their area.

Because The Source and Compuserve feature themselves as inexpensive computer networks dedicated to the consumer/hobbyist, they both offer a wide range of possible choices. They do, indeed, attempt to be all things to all people. In most instances they succeed. Many consumers find that the vast array of information which can be searched through on these two Utilities fulfills all of their expectations.

But, the professional and the businessperson may reach a point where more information than the utilities provide is needed. At that point, it is time to take a look at the offerings of two other well-known networks. Though they are both more expensive than The Source and Compuserve, Dialog and the Dow Jones News/Retrieval System give access to a wide range of highly specialized data.

Dialog is the more costly of the two in its full-blown, most powerful version. As our following charts detail, it offers databases in various professional and business areas. Costs, however, can range from $10/hour to well over $100/hour in connect time, and there is often a charge for each individual record found during a search. A typical example of this is Dialog's CHEMSEARCH database. CHEMSEARCH contains all new chemical substances cited in the magazine *Chemical*

Abstracts. The charge to enter this database is $130/hour, plus $0.08/record displayed onscreen.

Needless to say, Dialog is not the type of computer network that most users take leisurely, telecommunicating "strolls" through. When you enter Dialog it is usually with a definite purpose in mind, and a very good idea as to how to achieve it.

Dialog has recently begun a subsidiary database called the Knowledge Index. The network has taken some of its more popular databases and put them online at a cost more competitive with the consumer utilities. But, the databases are a stripped-down version of what is available in the regular area. These databases, as the name of the service implies, are just indexes to reference works which may be ordered.

We recommend that a potential Dialog user begin with the Knowledge Index. It will allow the user to become familiar with the type of searching commands used in Dialog. Developing facility with these commands takes time—indeed, Dialog runs seminars in various cities on how to search efficiently. The seminars are expensive enough that, for the same money, the Knowledge Index could be used for enough hours to get the general idea of the system.

The Dow Jones News/Retrieval System is another network alternative. The depth of business databases available there (see our charts) make it a must for the serious investor or financier. The cost is more than either The Source or Compuserve, but less than Dialog.

One of the best features of the Dow Jones service is the software support available for many of the most popular microcomputers. As detailed below, this software will allow you to automate much of your usual accessing and searching—and this has the effect of making the system more cost-efficient.

Whether a telecommunicator needs Dialog or Dow Jones depends on their need for the specialized data offered there. Dialog offers up to $100 worth of connect time free to first-time users, and the cost of Dow Jones for a few hours can be kept to a reasonable level. So, if one is not absolutely certain whether or not these services are needed—it is certainly possible to audition them.

Many people also wonder whether they should choose The Source or Compuserve. As the chart shows, there are more similarities than differences in many areas that each covers. But, we have found that the main differences lie in what we consider to be one of the most important areas — the interactive area. The mail facilities on The Source, for example, are more powerful than those on Compuserve. But Compuserve's CB Simulator allows for many people to conference at the same time instead of The Source's one-on-one CHAT feature. And, the Special Interest Groups (SIGs) on Compuserve are, in our opinion, far broader than The Source's Participate program.

The following charts compare the networks available to the average consumer.

Chart One
Compuserve and The Source Comparison

	COMPUSERVE	THE SOURCE
INTERACTIVE USES	**CONVERSATION** "CB Simulator" This is one of the most popular "places" on any of the information utilities. Users from all over the country talk in multi-layered conversations.	**CONVERSATION** "Chat" A communications channel that allows two users to communicate via their terminals or computers.
	MAIL "Email" This electronic mail service allows for quick delivery of mail between any two account holders. Not as powerful as SMAIL on The Source, but easier to learn and use.	**MAIL** "Smail" Sometimes referred to as Source Mail, this is the electronic mail backbone of The Source. It is a very powerful system that can also interface with a user's stored files.
	CONFERENCES "SIGs" The Special Interest Groups (SIGs) now cover over fifty areas of interest—from various computers to games and recipes. Each SIG is a structured bulletin-board where messages may be left, retrieved and online conferences may be held.	**CONFERENCES** "Participate" This tree-style conference structure allows for a free-flowing exchange of ideas. A user adds his own comments to existing conferences, or may start his own. Reachable with the command PARTI from Command Level.
	BULLETIN-BOARD "Bullet" The National Bulletin Board where items may be advertised for sale, as needed, or general notices may be posted. Much of this area's usefulness has been pre-empted by the SIGs, but it still has much activity. Reachable with the command R BULLET from Command Level.	**BULLETIN-BOARD** "Post" This national bulletin board contains categories to cover just about any interest. Retrievals are done by searching each category by keyword.

SHARED FILES
"Public Access"
In this huge database is a pot-pourri of free software and text files donated by members throughout the country to be shared with others. Reachable from Command Level with the R ACCESS command.

SHARED FILES
"Public Area"
The User Publishing feature on The Source allows every user a chance to be an Information Provider. Personal files may be opened to others—and the provider receives a "cut" of the connect time generated. But the user must also pay for storage space of his own files.

EXCHANGE
The Software Exchange (SOFTEX) allows users who want to sell programs to do so on a computer to computer basis.

MAILGRAM
Connecting to the Western Union Mailgram database allows you to send messages from your computer to anyone in the country—even if the ad-dressee has no computer or terminal. You can also batch mail to a list of names.

SERVICES

SHOP-AT-HOME
This electronic shopping ser-vice provides the opportunity to purchase via your computer and modem. Items include clothing, cookware, ap-pliances. Many discounts are available.

SHOP-AT HOME
Very similar to the shopping service on Compuserve. This too includes all types of ap-pliances, clothing, etc., with many discounts.

HOME BANKING
Enables subscribers to pay monthly bills and receive daily information. This includes summaries of deposits, checks, current balance, fund transfers and such.

HOME BANKING
At this writing, there are pro-grams to help you with bank-ing (see Financial Category) but as yet no online banks.

ENTERTAINMENT
Various forms of music, theatre, movie and television

ENTERTAINMENT
Featured in various areas are movie, music, theatre and TV

reviews are featured in various areas of the boards. Some of these areas are databases prepared for major cities. Other types of reviews may be found on the many Special Interest Groups.

reviews. The Mobile Restaurant Guide provides national restaurant reviews.

TRAVEL
Airline schedules, travel Special Interest Group.

TRAVEL
Airline schedules, travel packages.

GAMES
Games range from the board game and card game styles to adventure. This utility also features some interactive games where players from all over the country are online and competing in realtime.

GAMES
Over seventy-five games cover many areas of interest from card and board to computer-style. No interactive games at present.

NEWS AND PUBLICATIONS

NEWS
National and international newswires that are searchable only by listed titles of story from menu-like structure.

NEWS
UPI News Service which contains more than forty categories of news and sports, features and editorials. The database may be searched by entering keywords.

MAGAZINES
Many consumer magazines publish their electronic editions here including Popular Science, Computers and Electronics, Better Homes and Gardens and others.

HOME AND FINANCE

FINANCIAL
Provides information similar to (but not as extensive) as that found in the Dow Jones News/Retrieval Service (see Chart Two).

FINANCIAL
Provides information similar to (but not as extensive) as that found in the Dow Jones News/Retrieval Service (see Chart Two).

CONSUMER AFFAIRS
News and advice on car care, saving energy, child care, children's tests and games, food preparation and recipes, rebates, photography, more.

CONSUMER AFFAIRS
News and advice on cars, on beauty and fashion, collecting, design and crafts, food preparation and recipes, toll-free numbers, more.

EDUCATION
Provides information on college planning, financial aid, SAT updates, and adult education of many types.

EDUCATION
Provides both college and financial aid information, instructional programs in language arts, foreign languages and mathematics.

READING AROUND
Compuserve offers an eclectic mix of informational resources. Some other areas include agriculture, architecture, civil service, fashion, science, computer manufacturers, new products.

READING AROUND
Among other areas, The Source offers more than 40 categories of information on computers, seven on math, twenty-four related to the sciences, more than twenty on statistical analysis.

Chart Two
Comparison of Dialog Information Service and Dow Jones News/Retrieval Service

	DIALOG	DOW JONES

OFFERINGS

AREAS
This utility covers many topics within the Sciences and Technology as well as the social sciences, arts and humanities. Business, law, medicine and current affairs are also available.

AREAS
News and features as published by the Wall Street Journal, Barron's and the Dow Jones News Service. Special electronic edition of the Journal offers abstracts of major editorials and front page stories along with the special summary columns "Heard on the Street" and "Abreast of the Market."

CONTENT
60,000 journals offered in 40 languages in the form of technical reports, descriptions of current research, dissertations, patents, conference proceedings, books, bibliographies, pamphlets, monographs, legislative documents, manuals, reviews, newspaper and magazine articles, extracts of

CONTENT
Dow Jones averages; current stock prices, options and corporate bonds; current day updates for U.S. Treasury issues, mutual funds, national OTC quotes, foreign bonds, and government securities; companies listed in NYSE, Midwest and Pacific Exchanges, and NASDAQ OTC-

corporate reports (on file with SEC), statistics, time series analysis, more.

traded companies; stock quotes from market floor (delay of 15 to 30 minutes); historical stock quotes for common and preferred stocks and warrants. News and features.

KNOWLEDGE INDEX
Less expensive subsidiary service allows connect charges competitive with The Source and Compuserve (ordinary charges for services above are much higher). This abbreviated service contains about 10,000 journals and publications. Simplified searching method. Complete articles are not on-line—user receives a reference and an abstract and may electronically order full text at additional cost.

SPECIAL
SERVICES

CORPORATE INFO
Directory-type information on businesses, associations and foundations, names of officers, addresses and business objectives. Detailed financial information on current and past performances.

MEDIA GENERAL
This financial database lists 3,200 companies and reports on revenues, earnings, dividends, price-earning ratios and stock price performance. All NYSE and AMEX companies, 800 major OTC companies, composite data on 180 industries.

DISCLOSURE II
Wide-ranging composite data not found in most annual reports. Information as up-dated and filed with SEC — includes corporate profiles, balance sheets, income statements, line-of-business segment data, five year trends, income and earnings per share.

OTHER
A Corporate Earnings
Estimator provides a consen-
sus from analysts at more
than 40 brokerage firms cover-
ing 2,400 companies.
Transcripts from PBS program
"Wall Street Week." Online
encyclopedia contains 28,000
entries.

CUSTOMER
SUPPORT

SUPPORT FEATURES
Subscribers may order hard
copies of any document. Request is processed by Dialog
and transferred to the document supplier who sends to
user and bills directly. A toll-
free line is provided for
assistance. Phone help can
guide user through the searching process. Dialog also provides free monthly clinics.

SOFTWARE
Most popular microcomputers
have available special software from the Dow Jones Service. Software will do following: Automatic retrieval of current securities prices, and
updating of portfolio values;
retrieval of historical securities
prices and current news items;
allows users to store, modify
and update 100 individual port-
folios with up to 50 stocks
each; tabulates current value,
gains or losses of each portfolio; provides menus to guide
user through 17 analysis
routines; generates report for
current values, net and per-
centage changes, and unreal-
ized gains and losses; graphs
historical performance of
selected stock; comparison
charting of information for
several companies.

8. Files on the Information Utilities

The last chapters detailed how the two national information utilities can be used, both passively and actively. In an active manner, the communication systems available on each of the utilities (such as CB on Compuserve or PARTI on The Source) make possible communication via the computer keyboard with any other person connected to the same utility. The various electronic mail systems can also be used to send and receive short messages. These methods, and others detailed in preceding chapters, by now have probably helped you to telecommunicate with computerists throughout the country.

A more passive, but supremely interesting, way to use the utilities was briefly described in our last chapter — how to use their videotex capability to garner all sorts of information from the literally thousands of subsidiary databases and information providers present on the utilities. By now you have probably read the news, perhaps figured a home mortgage, or even gotten a new recipe.

But all of that, as exciting — and as useful — as it is will prove secondary to the experience of using these utilities in conjunction with your home computer to not only eliminate distance but to do away with paperwork as well.

This chapter and the next will explain how you can store large *files* of information in your own filespace on the information utilities. Once these files — which can be anything from company reports or short articles to tables of tabulated figures — are stored on the utilities you can revise them, send them to other users, use them as reference, or anything that you could do with them if they were on paper. Except they are not on paper, so access is instantaneous, and changes can be made almost as quickly as they are decided upon.

There are two ways of originating and storing files on the utilities.

The first way is to originate them while connected to the utility. This is called on-line composition. As you will see, these files may be originated simply by typing them in from the keyboard. They can also be originated by programs running on the utilities.

The other way to originate a file is called off-line composition. That is when your own computer is used to generate a file while not connected to the utility's computers. In many ways, as we will see in the following chapter, this is the most powerful method.

But in this chapter, let us first understand just what a file is. Then we will generate a few files on the utilities and store them. We will access the files, change them, and begin to see that paperwork no longer requires paper.

FILES

The ASCII code which was discussed earlier in this book (chapter 1) enables us to send the characters and numbers that make up our language from one computer to another. Because the letter A can be represented in a binary code of ones and zeroes, it can be encoded as audio tones by your computer's modem and sent along the telephone lines. And, your modem can decode these tones from another computer and display them as letters.

But computers not only speak in ones and zeroes, they use the same binary language in their memory.

A letter of the alphabet—as an ASCII code—can be sent as tones on a phone line or it can be stored as magnetic impulses on a computer disk. Because the ASCII code enables information to be treated as computer data, all the high points of computer technology can be accessed from your own home.

Keep in mind that there is no conversion step to go through or to oversee. Your modem and computer instantly translate all letters and numbers to ASCII. A company report, which to you reads like a collection of facts and figures delineated by adjectives and nouns, verbs and convoluted prose, is just a very long string of ones and zeroes to the computers you are working with.

This long string of ASCII codes is referred to as a text file. If such a text file is decoded under the ASCII convention it would appear in readable numbers and letters. Just about all the file types we will be using can be referred to as text files. (Later we will discuss something called a binary file.)

The two information utilities treat the concept of files in slightly different ways. In the rest of this chapter we will look at how files are used on each—and how you can apply these concepts to your own working life. In the next chapter we will build on this knowledge and show how your personal computer allows you to maximize this power.

FILES ON COMPUSERVE

We promised in our chapter on Electronic Mail (Chapter 5) that at an appropriate time, we would soon discuss the mysteries of "FILGE." This is the time, because FILGE, the "File Editor and Generator Program," is how Compuserve allows you to originate, store and change text files.

If you have sent electronic mail (EMAIL) on Compuserve or if you have composed a notice or other message on the National Bulletin Board area—you have already used the FILGE editor. FILGE is a simple, line-based, word-processing program. It allows you to compose your own text files and to store them on disk space rented from the information utility.

Because FILGE is a program it has its own command set. FILGE commands are preceded by a slash (/) and are the first and only entries on a line. Any letters not preceded by a slash will be treated as text by the FILGE program. The command set is very easy to use, and most of the commands are mnemonic in that they are often the first letter of the command word they represent. The basics of FILGEing a file can be learned in a very few minutes.

To get into FILGE requires that you enter the Personal Computing area of the Compuserve Information Service. This is the Personal Computing choice from the top menu. When it is selected, a warning will be displayed that you are about to enter that area. The warning is there because you are about to leave all the friendly menus behind. When return is hit after that message the letters "OK" will be displayed. The OK-prompt, as it is called, signifies that you are in Command Mode.

You can get back to the DISPLA area of Compuserve and all of its menus any time you see the OK prompt just by typing the command: R DISPLA.

Now, to investigate the possibilities of FILGE, let's step through the process of originating and storing our first file.

From the OK-prompt, type FILGE MYFILE.TXT and hit return. In a moment or so, the Compuserve computer will respond with this message:

New File MYFILE.TXT created — ready

You have now entered the FILGE program. From this point on whatever you type will be interpreted as either part of the text file to be stored or as a command to FILGE on how to manipulate that text.

When FILGE MYFILE.TXT was typed it started the FILGE program and assigned the name of MYFILE.TXT to the text file to be entered. A file can be named whatever you like within limitations. The limitations are that the main name must be six letters or less. This can be followed by a period and an extension of three letters. The main name of MYFILE in our example is modified by the TXT extension which indicates that it is a text file.

Now that FILGE is ready, just type in a few practice lines. When you are done use the command /EX at the beginning of a line and this will automatically save the file to disk and end the FILGE process. Follow along with this example:

OK
FILGE MYFILE.TXT
New File MYFILE.TXT created — ready

Entering a file into the FILGE editor
is as easy as typing. Just type line by
line and hit return after each line. When
done, don't forget the "/EX" command.

/EX

OK

Note how the /EX command returned you to the OK-prompt and to Command Mode. At this point, let's see if FILGE worked and if the new file has been proper-ly stored on the disk. To do this we will use a new command called CAT. This com-mand lets you look at the "catalog" or the table of contents as to what files are stored in your disk space. Below we see in the example that just one file (MYFILE.TXT) is stored in this user area of disk space:

OK

CAT
MYFILE.TXT

Any text file in your area can be typed with the TYP command. Here is how the MYFILE.TXT text file you entered previously would be read:

OK

TYP MYFILE.TXT
Entering a file into the FILE editor
is as easy as typing. Just type line by
line and hit return after each line. When
done, don't forget the "/EX" command.

OK

MYFILE.TXT has been successfully stored in your disk area. It can be typed out fully, as above, or you can return to FILGE and display only a portion of the file, change a word here or there, append more information, delete lines, or make any sort of changes. To see how this works, let's get back into FILGE and into our example file.

OK
FILGE MYFILE.TXT
File MYFILE.TXT — ready

This time the message was that your file is ready—not that a new file has been created because MYFILE.TXT already exists, as we saw when we entered the CAT command. You are now in FILGE and ready to display and edit your file.

FILGE expects the use of certain commands. Let's first look at the commands which allow for the display of text. Then we'll look at some of the commands that allow it to be changed.

The commands used to display text work on a line-by-line basis. It is convenient to think of the lines arranged in your file one after the other just as they would be on a printed page. Choose what line you want to display (or edit). Try to envision these commands as moving an invisible line-pointer up and down throughout the file.

The command /T, for example, moves the invisible line-pointer to the top or the first line in your file. The command /P for print may be followed by a number. If /P3 were typed, the next three lines in the text would be displayed. If you had previously used the command /T to move to the top of the file, then the /P command will display lines from the beginning of your file.

The /N for Next command moves the line-pointer back and forth within the file without bothering to display the lines. Follow the command with a number which tells the number of lines to move down or up in the file. For example, /N3 moves the pointer three lines down into your text file. Remember that the /N command simply moves the pointer, only the /P command will actually display the text which the pointer is on. A negative number moves up, toward the beginning of the file. A /N-3 would move three lines back.

The /N command can be followed by the Escape key which would have the same effect as following that /N (Next) Command with a /P (Print) command—the line would be displayed.

The /L for Locate command will let you return to any word or letter-grouping (referred to as a *string*) anywhere in the file following where the line-pointer is. The format is /L/(string). To find the word "computer" go to the top of the file and enter /L/computer.

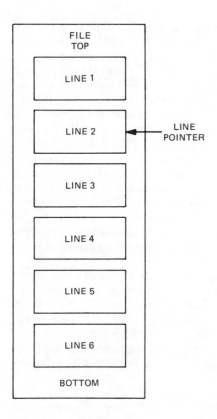

FILE
TOP

LINE 1

LINE 2 ← LINE POINTER

LINE 3

LINE 4

LINE 5

LINE 6

BOTTOM

Editor programs on the information utilities are line-based in nature. Think of the file you are writing as a "stack" of lines. The first line typed in is at the top of the file and more lines get added to the bottom as you continue writing. By moving a "line-pointer" you point at the line you want to change or read.

And, just as /T will position you back at the top or beginning of the file; the /B command will position you at the bottom or end.

Here's a quick look at how some of these commands would work on the sample MYFILE.TXT files:

OK
FILGE MYFILE.TXT
File MYFILE.TXT — ready

/T
/P3
Entering a file into the FILGE editor
is as easy as typing. Just type line by
line and hit return after each line. When
/N-2
/P

Entering a file into the FILE editor
/N2
/P
line and hit return after each line. When
/EX

First we got into FILGE and were told the file was ready for editing. The /T was entered more as a safe-guard than anything else. Ordinarily, a file is automatically entered at the top. The /P3 displayed the first three lines of the file from the top. The /N-2 moved our pointer back two lines and the /P displayed the line that command put us on (we could have just typed /N-2 followed by the escape key). We then moved the pointer two lines ahead, displayed the line and exited the file.

Now you can look back at your files, read any section of them at will, even automatically find words and phrases within them. But, without a way to change those files, all of this displaying would be empty play. The FILGE editor offers a number of commands that you may use to manipulate the text within your files.

The /C for Change command is very useful. This command will, in the current line, replace any string with any other string. The format is /C/(oldstring)/ (newstring). Let's say you have a line in your file that reads "This is my comptr." You would position the pointer to that line and display it. Then you would type "/C/comptr/computer." When displayed again the line would read, "This is my computer." (As an option you can erase the old string simply by leaving the new string choice empty.)

The /A for Append command will let you add text to the end of a line. If you have a line which reads, "The quality of mercy" then typing the command "/A/ is not strained" will result in "The quality of mercy is not strained." Note that a space should be left between the second slash and the beginning of the new phrase. Otherwise the words "mercy" and "is" would have run together.

To err is human, and computers allow for this with commands such as /D for Delete. Using /D erases the current line, /D5 would erase five lines including the current one.

The FILGE program may also be used to examine and edit text files generated by other programs within Compuserve. When the EMAIL program prompts you to file a message—it is filed in your file-space as a FILGE-readable text file.

Files in your disk area can be erased at any time using the DEL for Delete Command. At Command Mode, typing DEL MYFILE.TXT would erase that file from storage.

The FILGE program is one of the easiest to learn. And, as you do more file handling it will become one of the most important on Compuserve.

FILES ON THE SOURCE

To enter files on The Source it is necessary to be in Command Mode (choice 4 on The Source Menu). The Source allows for a wide number of versatile commands involving text files.

Entering a file on The Source is similar to Compuserve. From Command Mode, the command ED allows text to be entered. Using this command presents an "input" prompt. From then on, just enter your text:

— > **ED**
INPUT
Entering text in The Source's "Input Mode" is
quick and easy. Files are named after they
are entered. And, you end your editing session
by pressing return twice.

EDIT
FILE TEST.TXT

Note that, after hitting return twice, you go from the Input Mode to the Edit Mode. At this point either use the editing commands detailed below or—as we did above—store the file onto disk. The storage is done by using the FILE Command followed by a name for your file; a name may be 32 characters long.

The LIST Command will display the titles of all files presently stored in your disk space. Once a file has been stored, it can be edited at any time. Entering the ED command followed by an existing file title will "open" that file for editing.

The present mode will always be denoted by the appearance of an Input or Edit prompt when you change Modes. It is possible to switch between modes with a carriage return.

The editing commands are divided into two categories. The first category of editing command is used to display text. The second set of commands allows for changes in the text.

The display commands allows for the choices of Current Line. Only the current line may be edited. Commands are entered as the only input on a line while in the Edit Mode.

T for Top moves you above the first line in your file; while B for Bottom moves you below the text. These commands are useful for entering text either at the beginning or the end of the file.

The P for Print Command will display a line or lines—and the last displayed line becomes your Current Line. The P Command may be used with a following number (such as P5), and that is the number of lines below and including the Current line which will be displayed. P* will display all following lines.

N for Next will find the next line of text and display it—but the Current Line remains unchanged.

U for Up moves a line or number of lines above the Current Line.

F for Find will search the beginning of each line below the Current Line for any word or initial letter. (Entering "F And" would find a line starting with "And".)

The W for Where and PO for Point commands allow the editor program to line-number the lines. Using these commands, the lines in the text file are assumed to be numbered sequentially. The editor will number them if "Mode Number" is typed as a command. (The line numbers are for reference only and are not saved as a part of the file.) PO for Point will make the numbered line named (such as PO45) the displayed line. W for Where will give the line number of the Current Line.

Here is a quick example of a few of these display commands and how they work using the file we entered above:

– > **ED TEST.TXT**
EDIT
N2
quick and easy. Files are named after they
T
N
Entering text in The Source's "Input Mode" is
P3
Entering text in The Source's "Input Mode" is
quick and easy. Files are named after they
are entered. And, you end your editing session

Note that the "EDIT" prompt appeared when we entered the ED command followed by the title of an existing file. From there the N2 Command displayed the second line in the file. Then the T Command moved to the top of the file so that the P3 command would display the first three lines in the file (moving the Current Line pointer to the last displayed line).

The Source editor uses a number of commands to allow for changing text. First go into the Edit Mode with the ED Command followed by s text file's title.

The C for Change Command allows you to correct text within the Current Line. If the displayed Current Line reads "This is a tst" then typing C/tst/test would produce "This is a test." This may also be used with a Global Option which will replace the word in as many lines as wanted throughout the text. For example, C/Frank Wrong/Frank Wright/30 would (from the beginning of the file) replace Mr. Wrong's name with Mr. Wright's on the first 30 lines. Specifying G* would change it throughout the file.

The A for Append Command will allow for adding text to the end of the Current Line.

The I for Insert Command allows the insertion of lines of text within the file. After typing I you may use any of the positioning commands to move to the Current Line before which the new line is to be inserted. Type I, then the commands to position the line-pointer, then type the new line.

Replace (R) works like Insert (I) but it deletes the Current Line. If entered alone it leaves a blank text line. Or, a new line of text can be entered. By tying "R This is a new line" it will delete the old line and replace it with the new.

D for Delete will delete a line. A number following it (D10) will delete that number of lines.

You can add new text at any time by pressing return twice to switch into the Input Mode.

The editor on The Source also makes possible some very powerful options to combine parts of files with other files. This is done using the UNLOAD, LOAD and DUNLOAD Commands.

The UNLOAD Command will save a section of the current file to a new file. First the display is used to set the line-pointer to the beginning of the section to be transfered. Then the UNLOAD Command is used, followed by a new file name, and that followed by the number of lines from the current file that you wish to transfer. The format is UNLOAD (New-Name) (number of Lines). Typing the Command "UNLOAD SECOND.TXT 50" would take the next 50 lines in the current file and save them as a new file with the title "SECOND.TXT."

The DUNLOAD Command does the same as UNLOAD with one important difference. If the DUNLOAD Command is used the part of the text that is transfered to a new file will be erased from the old file.

The LOAD Command works in the opposite way from UNLOAD. LOAD allows for inserting a previously originated file into the presently edited file. The old file will be loaded into the new file following the Current Line. Enter the command as: "LOAD (Old Name)".

The FILE Command saves all of the modifications and returns to the Command Mode.

Other commands allow such things as sorting files and setting tabular formats. There is even a spelling checker available. However, as we will see in the following chapter, it may be more economical to do such things in other ways.

Many areas on The Source are set up so as to work with the files maintained in your disk area. The SMAIL program is a prime example.

During the electronic mail program (SMAIL) you may use the SAVE and FILE Commands at the "Disposition:" prompt. (See Chapter 5 on electronic mail.) The SAVE Command simply saves the letter being read as a text file. But the FILE Command allows storage of letters by category in a file called MAIL.FILE in your disk space. The Command structure is FILE (category)—use whatever categories best fit your needs.

Once your categories are set up with the FILE Command you can read or scan the letters there at any time. For instance, with a MAIL.FILE having the subcategory of CLIENTS you could type (at the Command Mode) MAIL READ FILE CLIENTS to reread any letters there. The SCAN Command works similarly.

And, when sending a letter in the mail area a special command can be used which will enter a file from your disk area into the text of your letter. This is the .LOAD Command. To use it you type .LOAD [filename] at the beginning of the line (don't forget the preceding period).

A CO file is a Command Output file. It contains commands used repetitively to reach a favorite area of the database. Just make a sequential text file of the command which would ordinarily be entered from the keyboard. The file is activated by typing CO [Filename] at the Command Mode prompt. A C—ID file is the same as a CO file but it automatically runs on logon. You must get into the editor with an ENTER C—ID Command.

The COMO command allows you to capture information as it scrolls onto the screen. Type COMO and a file title at Command Mode. Now use the Break key to get back to Command Mode. At this point everything you see on your screen will be saved into the file until COMO -E is typed. Be careful. If the COMO -E Command is forgotten there may be charges to store a file you never wanted.

All files can be erased from storage with the DEL for Delete Command. DEL PRACTICE, typed at Command Mode, would delete the file titled PRACTICE.

LETTING OTHERS USE YOUR FILES

Both The Source and Compuserve allow you to share your files with others. In this manner you can share information with others in your company, in your club, with friends, or with the general public.

Files are saved in a secure manner. Unless you say otherwise, you are the only person who can read them. You are guaranteed a level of privacy. Information which is yours—is yours. But to share information, this automatic privacy must be defeated.

SHARING ON COMPUSERVE

Sharing files on Compuserve can be done in one or both of two ways. With another person who is on the same network computer you can share files directly. Or, you can share with everyone on the network using a special program.

To share directly with another person involves lowering the privacy level of a file. This is referred to as the file's Protection Level. The protection level of a file is shown if you use the DIR for Directory Command. A typical display of such a command on a disk area which contained two user files might be:

OK

DIR

MYFILE.TXT	3200	10-Oct-82	(4)
FIRST.TXT	3200	12-Oct-82	(4)

The name of each file is followed by the number of "blocks" of storage it is using on the disk, by the date the file was last accessed and, in parentheses, the protection level of the file.

Level 4 is the highest level of protection. Only the owner of the file may read, write or (if it is a program file as opposed to a text file) execute it. A Level 2 protection would enable others to read your file. A Level 1 protection would even allow others to change it. (Level 3 is not used.)

First thing necessary to share a Compuserve file is to lower the file's protection level. This is done with the PRO for Protection Command. To lower a file titled MYFILE.TXT to a protection level of 2 would require typing at Command Mode the command: PRO MYFILE.TXT (2).

Besides having to individually lower the protection level of each file to be shared, there is one other safeguard to be dealt with. You will have to, only once, lower the overall protection level of your disk space or even low-protection level files cannot be accessed by others.

The disk area's protection level is lowered with the PRO# Command. The standard Protection Level on an area is 700—where no one but the owner can get in. Level 777 would give everyone complete control over any files for which individual protection has been lowered. Level 755 would allow others to read files which have been specified individually and to see the titles of even your protected files (but not read them). Level 744 would allow others to read your unprotected files and not see even the names of the others.

Typing PRO#744 would change your protection level. Then you would use the PRO command on each file in your area to decide if it could be accessed from the outside.

You can access another person's file area by appending that person's account number to a file command. For instance, typing DIR[70003,273] would display the Directory of 70003,273's disk area if the protection level as set by 70003,273 allowed. You could then look at a file whose protection level was low enough simply by using the TYP Command with the account number appended. Note that the left and right square brackets are mandatory.

However, all of the above will only let people who are running on the same host computer share information. Compuserve also has a program called Public Access which will share your text files with anyone on the network, with no regard to what host computer they are on.

First place the file in your directory, stored there via the FILGE editor. Then type R ACCESS. At the program's prompt use the SUBmit command to tell which text file you are submitting. Typing SUB MYFILE.TXT would place the MYFILE.TXT text file into Public Access. After the system picks up on the submission (about 24 hours) anyone can see the file name with a CAT command from within R ACCESS. You could, however, have typed SUB MYFILE.TXT/INV; the INV is called a modifier and this "Invisible" modifier means that only someone who knows the file exists can read it — its title won't show on a CATalog request.

In the R ACCESS area, files may be read with the TYP command. If you have a special program available from Compuserve (called the Executive program) you will even be able to download programs directly into your computer's memory. (If you don't have the program, you can still do this as we will see next chapter.)

SHARING ON THE SOURCE

The Source too allows going public with your files. As we did on Compuserve, here on The Source we will first have to let the system know that files may be accessed from our disk area by other people, and then we will have to specify exactly which files may be accessed.

First use the PASSWD program to modify your password. Type PASSWD from Command Level. Enter your old password when asked. For the new password, enter the old password but followed by a comma. If your password was "TERMINAL" then "TERMINAL," would be your new one. (This, by the way, does not change how you enter your password when signing onto the system. Do not type the comma when the system asks for your password.)

The next thing to do is to use the PROTEC Command on every file that you want to share. There are eight levels of protection on The Source:

	DELETE FILE	CHANGE FILE	READ FILE
Level 0	No	No	No
Level 1	No	No	Yes
Level 2	No	Yes	No
Level 3	No	Yes	Yes
Level 4	Yes	No	No
Level 5	Yes	No	Yes
Level 6	Yes	Yes	No
Level 7	Yes	Yes	Yes

Each file should be assigned two protection levels. The first level applies to you and the second level applies to everyone else. Setting PROTEC SAMPLE.FILE

7 1 would let you have protection level 7 (complete control) and would allow other users level 1 access (read only).

The FILES Command can be used to help keep track of what levels all your various files have assigned to them, as well as for finding other information. Typing FILES at Command Level will display a line of information for each file in your disk area. A typical line might be:

SAMPLE.FILE   003   11/19   15:01   ASC DWRR

This shows that the file's title is SAMPLE.FILE, that it occupies three blocks of storage space, that it was last modified on November 19 at 3:01 pm, that it is an ASCII text file, and that the owner can delete, write and read it—and the public can only read it.

You can read files that are made public by using the TY Command along with the account and system number of the other person's disk area. Entering the command: TY (13)DM0090 SAMPLE.FILE would enter your request to read the file titled SAMPLE.FILE in DM0090's disk space on System 13. Of course, DM0090 would have had to lower his disk area and file protection levels for your request to be fulfilled.

SUMMARY

The file structures on the information utilities allow storage and retrieval of information in the form of text files. You can also originate files or change the contents of files using the editing programs. Files can be made completely private or accessible to the public at varying levels of protection.

While the line editors on the utilities are good — there are better ways to go about originating your files. You'll likely always use the editors to touch things up. But in the next chapter we will show how to use your personal computer to take full advantage of the storage capabilities of the information utilities—while keeping the cost of on-line storage in line!

9. Uploading and Downloading to the Utilities

Just as radio can be a two-way means of communication, your computer can also be used as a communications tool which—in many ways—surpasses the capabilities of any other medium. Your computer is the key to the concept of the decentralized office as well as being a gatherer and storer of the world of information contained within the wide realm of telecommunications.

Your computer, as we have seen, can directly access hundreds of public access message systems (PAMS) as well as the thousands of databases and services on the information utilities. But, because you are using a computer which can be equipped with powerful software—you can address the utilities with even greater effectiveness.

As we have used them, the Electronic Mail services of the utilities have opened up new vistas in instant communications. But there are ways to use these services which will provide the same ability and dramatically reduce costs. We have shown how to read screens and screens worth of information from the utilities—now we'll see how we can save that information to be used at our leisure. We have seen how to share private files of information with people throughout the country—now we'll see how to efficiently originate such files and how to work best with files from other people.

The secret lies in mastering the art of uploading and downloading.

Just as a CB radio will transmit and receive messages, your computer can transmit and receive data. But your computer can receive information, store it automatically, allow you to study and modify the information, and then effortlessly retransmit it to other computerists as often as you might wish. The process of retrieving information, and storing it in your computer's RAM or disk memory is called downloading. The reverse process, of sending information from your personal computer's RAM or disk, is called uploading.

Let's recall how we have used the Electronic Mail on Compuserve and then see how the concept of uploading and downloading makes it far better, cheaper and more efficient. First, a quick summary and then we will examine, in depth, how to better our already good electronic mail capabilities—and then we will look at many other applications.

When entering the send portion of the electronic mail or EMAIL program of Compuserve (GO EMA-4) you are first asked to supply the program with the text of the letter to be sent. We have so far just composed the letter by typing at the keyboard. But, there is another and more efficient way to originate such text—to send the file from your disk storage area.

At first glance, this may seem a "six of one and half a dozen of another" choice. After all, so far the only way we have seen to enter a file into our disk storage area also involves typing it in, listing, and modifying it with the commands discussed in the previous chapter.

But, whenever you are on one of the information utilities, the clock is running. Even if you are a touch typist, entering a long and involved letter can cost far more in connect time than even the most expensive overnight mail.

The solution is to compose your letter on your own computer before you connect to the utility, and before the per-hourly rates begin to mount up. Only then, when the letter is ready to go, do you connect to the utility.

You then upload the letter from your computer directly into your disk storage area on the utility. This takes only a very small fraction of the connect-time that composing the letter would have required. And, once the letter is in your storage area as a disk file you simply use thi electronic mail program to send the file to someone else.

Uploading, with most computer and software arrangements, is an easy-to-learn, very logical and orderly process. Here's how the uploading process works in general—and in specific for our do-it-yourself system.

Uploading And
The Word Processor

Before you can upload, you need a file to send. One of the ways to do this is to originate your own file. It can be a letter, a business memorandum, a table of numbers, an inventory list—just as long as it can be stored on your computer's floppy disk memory as an ASCII text file.

If you were still in the age of paper and post offices you would type your letter on a typewriter—and use a lot of opaquing fluid or correction paper. In today's world of instant and perfectly composed telecommunication, you word process a file.

We have already had a little experience with word processing. The file editors that we have investigated on both The Source and on Compuserve are primitive word processors. Like all such word processors (or "editors" as we have referred to them previously) they allow entry of text by typing on your keyboard and then allow viewing of the text on your screen. They also allow changes, deletions, and insertions to be made before saving a final version.

But compared to even the least ambitious word processor available for personal computers, the file editors on the utilities are difficult to use, slow, and not nearly as versatile.

The editors we have seen on the utilities are known as line-based editors. Moving the line-pointer to the line wanted allows modifications of a file on a line-by-line basis.

Most word processing programs for personal computers are character-based editors. In a character-based editor the text is seen as it is typed on the screen. At any time a word or phrase in the document can be changed without having to think of the document in terms of line numbers. Instead, a visible cursor is moved right on top of the word or phrase to be changed, and the changes are entered.

This makes a character-based editor far faster to use than a line-based one. The document is saved as a file on your computer's floppy disk. Later, the document can be recalled from the floppy disk storage and re-edited or printed on an accessory printer.

The monitor screen works as if it is a window onto your text. The text can be "scrolled" up and down behind this screen window so as to bring any area of the text into view. Once you see what is to be changed, other commands allow moving the place-marker (called a "cursor") to the area to be corrected. It's that fast, and that easy.

The information utilities cannot provide this form of editing because of the fact that so many different brands of computers connect to them. While all the computers do share the ASCII code as far as the letters and numbers of the alphabet—in other areas, such as the special control codes that control the way information is displayed on a screen, the computers are still incompatible.

If a programmer writes a word processor program for a certain computer—he knows the kind of control code that computer expects to cause it to, for example, clear the screen to a blank. But the Apple requires a different clear-the-screen code from the TRS-80, which in turn is different from the Atari 800, and so on. And terminals which use paper printouts have to be handled differently from screen-based computers and terminals.

So, the line-by-line system of word processing, which does not require any special control codes be sent to the computer or terminal from the utility, is the only viable method of word processing on a system such as The Source or Compuserve. But, by itself, this form of word processing does not even begin to give you the power and versatility of a word processing program on your own computer.

While we cannot, in this book, discuss all the aspects of word processing, we can certainly recommend ways to choose and use a word processor for telecommunication applications.

The most important thing in a word processor for this use is that the files the word processor saves to disk be compatible with the standard ASCII codes used in telecommunication. Many word processors do not save their documents as ASCII text files. Instead, they store documents as binary-encoded files. They use their own encoding scheme, which may enable them to retrieve their own style files faster from disk and to store them in less disk space. These are laudable goals for a person who never expects to send information via modem—but for us telecommunicators such files are useless. To stress the point: Your word processor must store your document files as normal ASCII text files.

Your disk drive reads and writes information using software which is called the Disk Operating System or DOS. The DOS is a machine-language program which is stored on every one of your master disks—a master disk being any which will run (or boot) on the computer by itself. Some word processors use a DOS which is different from the DOS your computer expects — and different from the DOS your terminal program runs under. The reason again, is usually speed of access to the document files. But, this too can cause telecommunication failure.

Your files must be on a disk that supports the same DOS as your terminal program. Otherwise, your terminal program will not be able to "read" the files produced by your word processor.

As an alternative to a full-fledged word processing setup, some terminal programs include a "stripped-down" word processor. Although these built-in editors are never quite as good as having a completely separate word processor they will still serve for many requirements.

A good word processor program can be expensive. If your terminal program includes an editor, then that will at least be enough to get you going. And, if the terminal program has the editor included then there is nothing to worry about regarding compatibility between your files and your terminal software.

The AE Pro terminal program (in our do-it-yourself system) does include an editor which features better word-processing capability than most other terminal programs. Later in this chapter we will take a look at it.

INTELLIGENT UPLOADS

Your terminal program should be capable of some "intelligence" during an upload. This means that the terminal program should be able to recognize any of the prompts that the program on the information utility is sending.

For instance, let's take an example of uploading a message on one of Compuserve's Special Interest Groups (SIGs).

The SIG software expects that you will be loading, or typing, in a message on a line by line basis. You type in a line of copy; then hit return. The utility may pause for a moment. Then it will display the next line number, followed by a colon. The colon (:) is considered a system prompt. Unless it has been displayed on the screen, the utility is not yet ready to accept information from your computer. You must wait for the prompt before beginning to enter information.

If you are at the keyboard typing, this presents no problem. But, if you are trying to upload a file from a disk, the whole system breaks down. Your computer sends a line automatically. The utility does not accept this line instantly. The utility expects the next line will not be sent until after the prompt. Your computer will, dumbly, keep sending the file even though the utility's computer is not receiving. The best that could happen would be for your message to be garbled and for lines to be lost — the worst is that the entire upload would not work.

This situation arises in many areas on all the utilities. Remember that these utilities are time-sharing systems. You are sharing the computer facilities with however many other users are also connected. No matter what you are doing, you have to "wait your turn." Sometimes this wait is quite short, but often enough the utility computer is simply not yet ready to give you your turn. You will often have to wait for a prompt before typing in information — or sending it by upload.

Make certain that your terminal program will know enough to stop and wait for any special prompt.

DOWNLOADING

Transmitting a file is only half the story; your computer must also be able to receive information and store it as a file. Half a connection is no way to communicate.

Many times while on the information utilities, or the Public Access Message Systems, you have seen words and data flow in, fill up your screen, and scroll up off the top of your screen into video oblivion. Even at the relatively slow speed of 300 baud, it is a quick and seemingly irreversible process.

Your terminal program must allow for a way to "save" this information from disappearing forever. Most terminal programs do this by means of a *capture buffer*.

A capture buffer saves all incoming information in your computer's RAM memory so that it may be either examined directly from RAM memory or stored on disk. Obviously, a capture buffer cannot be considered a bottomless bucket in which to store a whole day's worth of heavy telecommunicating. A personal computer's RAM memory is limited. Your terminal program can only use as much memory for temporary storage of captured text as there is memory beyond what the program itself uses to run.

In telecommunications work your computer should be equipped with as much memory as it can utilize. Lately, there have been a number of memory cards available from many suppliers for the various computers. These cards are capable of taking a computer far beyond its addressable memory limitations. Using these cards, it is possible to have almost a megabyte or more of RAM memory in computers that were, perhaps, designed for 64K.

To use these cards with most telecommunications packages requires that they be supplied with "pseudo-disk drive" software. As an example, think of an Apple II equipped with 64K of RAM on the motherboard and a plug-in card with 128K of RAM. As far as the Apple's CPU is concerned, 64K is a top limit. Without supporting software, that 128K extra is inaccessible.

Let's suppose that this system is running a terminal program with the capture buffer turned on. The text scrolling in on the screen is being copied and stored in the computer's 64K of RAM memory. Gradually, that 64K begins to fill up.

At the point the 64K is filled, the terminal program will do one of two things. It will either (through a bell or other warning) alert the computerist that the memory is filling—or it will automatically "dump" the memory to disk. During a memory dump, the information that has been captured is saved onto disk as an ASCII text file. In a good terminal program, this saving process will be automatic. It can also be initiated at anytime by the operator.

If you are running a memory card it should be treated as a pseudo-disk—a very fast disk drive. It takes a good length of time to write a very long file to a mechanical disk drive. The write head must seek track after track, the mechanism must spin, all of these short delays can add up to a considerable length of time. But with a memory card with the right software this dump process only takes a very few seconds, since there are no mechanical parts.

Two things, however, must be remembered if you use a memory card. First, the card itself will likely be unusable without a pseudo-disk program for it that runs under the same DOS (Disk Operating System) as your terminal program.

Secondly, never forget that information stored in RAM memory—whether the RAM is on the card or on the computer's main board—will only last for as long as the computer is turned on. As soon as the computer is turned off or power is interrupted, everything stored in RAM whether on the main board or a plug-in card is lost.

Obviously, your downloading operations must constantly be monitored. Make certain that all very important information is consistently saved to a real disk drive. You will probably not remember this until the first time a few hours work is lost—then, like everyone else that this sooner or later happens to, you will never forget again.

YOUR FIRST UPLOAD/DOWNLOAD

Let's take a step-by-step look at how you might typically use your up- and down-loading capabilities while on one of the information utilities. We will illustrate this with the command set supported by our do-it-yourself system and the AE Pro terminal software, but no matter what system is being used the basic tenets should hold true.

1) Dial onto one of the information utilities. For purposes of illustration, we will assume that there is electronic mail waiting for you. Scan your mail (if on The Source) or get to the mail-waiting menu (on Compuserve) and choose which letter you want to read and capture.

2) Turn your capture buffer on. Using AE Pro, the command sequence for this is: control-Q,R. (This command is one of the ones visible on AE Pro's Primary Command Menu shown when you enter control-Q,1.) AE Pro will prompt you that "Copy is ON"; other terminal programs will give some similar indication. At this point, everything happening on screen — what you type as well as what is scrolling in — is being saved to RAM memory.

3) Type the command to read the letter you've chosen (see our Electronic Mail chapter for full details on this). The letter will scroll normally on the screen.

4) Turn your capture buffer off. The AE Pro command sequence of control-Q,R toggles the buffer from on to off. AE Pro will tell you copy is off, other programs will give a similar indication.

5) You may optionally view the buffer. This is a check that the information wanted really is in your computer's memory. AE Pro expects the command sequence control-Q,V. Entering this View Buffer command will cause the captured memory to scroll on the screen (far quicker than at the 300 Baud rate it was originally sent over the phone lines).

6) Now dump it to disk. AE Pro wants a control-Q,W (for Write) command along with a name for the new ASCII file the information will be stored as.

7) Now sign off the information utility. (On Compuserve you will first have to answer that the letter you are reading should be deleted from your mailbox.)

8) Use your word processor or the editor portion of your terminal software to examine the letter now stored as a text file on your computer's floppy disk. AE Pro has a handy built-in editor. The Y command loads the editor portion of the AE Pro software. Once in the editor the .G command (for Get) followed by the name you gave the file will load that file. The .L command will List it line by line. Use the .X command to get out of the editor and then you can use the control-Q,P command to turn on your printer and the control-Q,J command to printout the file.

Obviously, various word processor and/or terminal program combinations will require differing commands. The main thing you want to do is to prove that the file you have captured is now accessible for changing, modifying or reading again with whatever editing or word processing facilities are being used.

9) Now compose a letter on your editor or word processor. Save it to disk as a text file under a new name. (With AE Pro, that calls for another trip to the editor—the Y command. Then use the .C command to Clear memory. Type in your letter. Use the .P command to Put it on disk.)

10) Dial onto and connect with one of the information utilities. On Compuserve get to Command Mode (the OK prompt). On The Source get to Command Level

(the – > prompt). As mentioned in the previous chapter, these are the areas from which you can enter files into your disk storage area on the utilities. But now, we will enter a file without typing it in and paying for superfluous connect time.

11) Open a file space on the utility (the FILGE [filename] command on Compuserve). When the utility prompts you to begin entering your file, use the upload capability of the terminal software to send the letter just composed. AE Pro requires the control-Q,S (for Send) command sequence. You will want to send line-by-line, with no special protocol or prompt character. For filename, supply the name of the text file on disk that you wish to upload. The file will be read off your computer's disk and sent to the utility just as if it were being typed in.

When transmission is complete, use the proper command sequence to close the utility file. (On Compuserve the /EX command will return to Command Mode. On The Source, it takes two carriage returns to get back to Command Level.)

12) Now send the file from your disk storage area by using the electronic mail facilities. (On Compuserve type GO EMA-4 and pick the choice of sending a message from your file space. On The Source use the .LOAD [filename] command from within the SMAIL editor to insert your file as the body of your letter.) Of course, if you want to send the same letter many times, just keep using the stored file rather than composing a new message for each addressee.

This process of up- and downloading can, of course, be used in many areas other than electronic mail. Where it will be used depends on your particular interests on the utilities.

If you plan to work with stocks and commodities, bonds and other financial materials, you will find this ability to capture information and work on it at your leisure to be invaluable. Rather than having information scroll by you and be lost — you will be able to store it on disk to go over at any time and with no further expense.

If you are one of the many people who uses the database facilities for researching news and events, you will find that the downloading capability will enable you to store vast quantities of information. Then, as we will be demonstrating later in this book, you will be able to use your own computer to search for items of special interest or to sort your data in a way that is easiest for you to use.

THE BEGINNINGS OF POWER

You have now, for the first time, truly used the power of computers and modems in a two-way telecommunication.

As we have seen, the national information utilities are much more than collections of databases and information. They are the very medium of transmission for today's most versatile means of communication — the personal computer.

These wide-spread computer networks allow you to communicate via computer with any other computer user in the country who is also connected to the network. We have seen in previous chapters how files may be shared through the network with others — and we now know inexpensive and quick techniques to put such files onto the networks and to take them off for our own use.

Yet, as interesting as these computer networks are, as powerful as they are, they are not the whole story.

In the next chapter we will take a look at how you can begin your own computer network, dedicated exclusively to your own needs and accessible to whatever sector of the public you designate—from your corporate structure, to your bowling team.

10. The Power of Two

The last few chapters have described the world of the information utilities, powerful mainframe computers that, through the process of time-sharing, allow thousands of personal computers to network together. But, there are other "networks" as small as just two computers connected together. And, sometimes, far more in the realm of information management can be achieved by setting up one of these private "networks" of your own.

There are various degrees of sophistication available for a private network, depending on your particular needs. Your computer can be remotely accessed from a second computer. The second computer can assume complete control over the first. Furthermore, this power can be reserved to yourself or to any number of callers who have your permission to access the system. Or, the system can be set up as a private messages-only network whose contents are protected by a series of passworded security levels. You can even go the other route and set up a combination message center and computer fully remote-controllable by a vast number of people.

Primary costs for any of these alternatives will involve only your investment in the necessary software. The more information processing which you can move onto your personal system, and off the utilities which charge for connect time, the greater will be your savings.

The simplest form of telecommunications network consists of two people just typing back and forth between two computers, whose modems are connected via the phone line. This is very easily done. The telephone connection is established normally, and the operators at each end of the connection switch to their modems at a half-duplex echo. (Half, rather than full, as screen output in this case is dependent on the computer placing it there as a key is hit rather than waiting for an echo.) Once the modem connection is established, the two operators can "talk" back and forth over their keyboards in pretty much the same manner as an old radioteletype-style connection.

While this is a lot of fun the first time you do it with a computerist friend—it quickly becomes apparent that no real treasure trove of information management possibilities has been uncovered. A two-way, typing-only connection is far less useful than just a simple voice communication. Typing is, of course, harder for most of us than simply speaking our sentences.

But there are many types of communication that you can do far better over a computer connection. Here follows a summary of all the basic ways in which you can use your personal computer as the nexus of your own small network, along with detailed descriptions of how each applies to our do-it-yourself System.

WITH A LITTLE BIT MORE

We have seen that no software is required for that most basic of telecommunications—computer to computer typing via modem. But, we have also seen that this is the least useful of all that can be done with a computer. Just a bit more software will begin to open wide the doorway of possibility. It doesn't take much to allow two computers to exchange stored text files, or even programs. And, if one computer is unattended, it is easy to run it using the manned computer as a remote terminal. With more powerful software you can use your personal computer as a veritable fount of information which can be accessed by your confidential business associates or close friends, by anyone who calls in, or just by yourself from another location.

Our particular do-it-yourself system is all ready to go as a remotely-accessible terminal. This is due to the firmware on the Hayes Micromodem II. (Firmware is a computer program which is stored in a ROM memory chip on the circuit board.) The firmware on the Micromodem II circuit board is referred to as the Micromodem Terminal program.

This Terminal program allows the Apple II to be telephoned —and controlled in a limited manner—by any other computer or dumb terminal calling through the use of a few, simple control-code commands. The amount of control which you will have, without any other supporting software, is minimal.

But, if you are away from your system—and you have either a portable terminal (such as the RCA VP-3501 mentioned in the introductory pages) or have access to another computer —you can call your Apple on the phone. Within limits, many of the same things can be done from your remote location as could be done right at the keyboard.

By remote control you will be able to type interactively, run simple programs, look at the titles of files on your disks, and so on. Let's take a quick look at this addition to the bare computing essentials; and then we'll see what more sophisticated functions can be accomplished by using some of the available software.

TERMINAL PROGRAM

The Micromodem II's Terminal program requires that the Apple be in a ready-to-be-called state before you leave its side. This is done by using the Apple II's IN# and PR# commands from its built-in BASIC language or control codes from its machine-language monitor mode.

Turn on the system as you would normally. The best way to do this is to simply boot the System Master disk, which will also load in the disk operating system (DOS). Then, at the Apple's BASIC prompt, type IN#2 and return, then PR#2 and return (the number is whatever slot your Micromodem is in; we will assume it is in slot #2). If you are in the Apple's monitor you can achieve the same result by typing 2 control-P return and 2 control-K return. In most cases, however, it will be far easier if you are in Basic rather than in monitor (The Apple's monitor issues an asterisk (*) prompt. If DOS has been loaded, return to Basic by typing 3DOG.)

Though there are no indications on-screen, The Apple is now looking for input to the Micromodem's Terminal program rather than to the keyboard. You can now go to another computer (or terminal) and call the line to which the Apple is connected. If someone were watching the Apple's screen while you called in, they would first see the words:

MICROMODEM II:RING

After the first ring, the Micromodem will answer the phone. Then it will turn on its tone carrier and display on the Apple screen:

MICROMODEM II:AWAIT CARR.

Finally, if it is answered by another computer's carrier tone in the next 30 seconds it will display:

MICROMODEM II:CONN.

It will now be connected and ready to go. On your remote screen, however, you will have seen none of this. The first indication that you have successfully called your Apple II will be when you see the Apple's prompt appear on your remote terminal's screen or display device. The prompts, of course, are the familiar] if you have an Apple II Plus, a > if you have the Apple II, and * if the Apple is in the monitor.

It doesn't matter if you are halfway around the world, you are now in control of your machine from your remote location. You should be able to catalog the disk drive and read the file titles. You could even type in a program and then save it to the remote disk system.

But you will quickly discover the limitations of this simple system. The Apple's graphics will not be available remotely — even if you are calling from another Apple. The modem has no way to communicate a graphics display in real-time basis. You will also be unable to use the port commands (such as PR#1) to, for example, turn on a printer without disconnecting the modem and hanging up on yourself. Nor will you be able to upload or download easily.

As the Micromodem's manual details, there are a very few control-code commands to help you out. The [control-S] stops the Apple's output, and is very useful if the information is scrolling by too quickly to be comfortably read. Any other key restarts output. Other commands such as [control-T] and [control-R] switch or toggle between the Micromodem's Terminal and Remote modes. The [control-Z] command allows you to hang up the Micromodem but leave it connected and listening to the phone for your next call. The [control-N] command is important if you are calling in from a remote terminal or computer which uses a printer as a display output device. The [control-N] supplies the linefeeds needed by some terminals and most printers, the delay at the end of a line to allow for a physical carrier return, and also allows use of the full width of the printer.

But the best part about the Terminal program is the way it invisibly helps other software that works in conjunction with the Micromodem II to access the full telecommunications ability of this system. We'll now see how some of these other, more complicated programs, really bring the remote capabilities of telecommunication to life.

SENDING FILES AND PROGRAMS

While we have already seen how to use the firmware Terminal program on the Micromodem II, such a simple program will have many drawbacks if used on a daily basis for complicated tasks. And, there are many things that cannot be done at all with this firmware program. For example, if accessing an Apple remotely, it would be difficult to transfer a long machine-language program. It would be equally difficult to be certain that a long inventory sent as a copied text file did not drop an important line or item. Even worse for some businesses would be the possibility that the competition could simply obtain the number of your Apple, call and access important information.

Part of the answer to these problems is to choose a terminal program which includes a number of useful subroutines for remote capability. The software normally used to access utilities and BBS computers should also be capable of running the Apple unattended—to answer the phone as well as dial it. And, once the phone has been answered, the software should promote the easy transfer of text files and programs from your personal computing system to whoever is authorized to call it. The word "authorization" is crucial. There should be some means, always, of protecting your system from a caller who wants to create telecommunications mischief. And, if possible, the software should support some widely recognized form of protocol error-checking.

Many terminal programs for the various personal computers have all or some of the above features. One of the reasons we chose the AE Pro terminal program to run on our do-it-yourself system is that this is one of the programs which have all the features mentioned. Let's take a look, for both example and practice (if you have our system), at using these types of features.

It should be very easy to put your terminal program into a remote mode. And, once so set up, the program should give some constant indication of its status. AE Pro is set up with a one-key command from its menu structure. When from local or connected mode, a plus-sign (+) is typed, the program goes into its unattended method of operation. The screen displays:

Calls received: 0
Waiting for ring . . .

The number of calls received is the number since the program session began. The number of rings it will wait for is one of the options which can be defined in the Install program. It is usually set to one ring. Hitting another plus-sign (+) would direct AE Pro not to bother waiting for a ring but just to pick up the phone and wait for a carrier—most useful if you are already connected on a voice line. (If the other carrier is already there when unattended operation is selected, the display will be bypassed and you will find yourself automatically connected.)

But, let's say that this is a normal unattended operation where AE Pro will be waiting for a call. In that case, the caller will be greeted with this prompt on his screen after he makes contact with your computer:

Entry:

AE Pro, at this point, is demanding that the caller type in a password. This, of course, is your guarantee that the information left on the computer system can only be picked up by people who have previously been given the password. Unless the caller types the right password, he will not gain access to your computer. And, there is no time for a lot of guessing. If one minute passes or 20 wrong characters have been entered and the password has still not been correctly typed in—AE Pro will hang up on the unauthorized call.

If the password is correctly given, AE Pro will then issue a special prompt to show it is waiting for remote commands. The prompt is a combination of the right parenthesis and greater-than sign which is:

) >

At this prompt, AE Pro will wait for one of six commands from the calling computer. Some of the commands depend on the calling device being a computer with software running; other commands will work with a dumb terminal calling in.

If, at the remote command prompt, a C is entered (for Copy) then AE Pro knows to first ask for a filename, and then to save all incoming data to that file. At this point, if the calling machine is a computer it can upload a file from its own disk storage. Or, if it is a terminal, the operator can type in a message. Either method requires that, at the end of the file, two control-C's can be sent to signal AE Pro to close the file, save the information thus far to disk, and return to the remote command prompt. This process can be aborted with two control-D's sent one after the other. This method is, of course, solely for the sending of typed-in text or information stored as ASCII text file characters.

The D for Directory command displays the catalog of file titles already stored on the disk. The V for View command will read an ASCII file from the remotely-accessed disk drive. This last is a good, visual check to see that the information sent during the Copy command sequence arrived and was saved in an unscrambled condition. There is an H for Help command which simply prints out a list of the on-letter commands the remote command prompt will accept.

The final two commands—S and R for Send and Receive files—are the two most powerful. These commands not only allow sending ASCII text files, they also make possible the transfer of BASIC programs and machine-language, binary files from one computer to the other. And, besides just sending and receiving the files some very sophisticated error-checking and auto-correcting is done as well. The catch is that the calling computer must be capable of supporting the same "Christensen Protocol" that AE Pro supports.

Ward Christensen is a computerist who has been credited by many as more or less inventing the concept of the electronic bulletin-board with his CBBS program. (See Appendix One for a list of the CBBS-style style of boards as well as the many other types.) Many electronic message systems which run under the CP/M operating system already make use of the Christiansen Protocols. Other operating systems, running on various computers, are now also beginning to use this means of transferring files. The AE Pro terminal package was one of the first to make these protocols available to the world of the Apple.

The protocols enable a file to be sent from one computer to another under a constant error-checking. The files are sent in sections called blocks. Each block can be automatically re-transmitted if an error needs to be corrected. The actual details of this would be of interest only to programmers (an excellent discussion can be found in the AE Pro manual) as it is all done automatically without any need for the computerist to do anything but feel confident that a transfer will work. To add to this confidence, AE Pro has one other command, called CRC, which is automatically called by the Christiansen protocols (but may also be called from AE Pro's Secondary Menu by the "/" command). This Cyclic Redundancy Check provides a number in response to a file name. If the number is the same for the original file and the transferred copy, then all has gone well.

Here is a quick step-by-step summary as to how to use these protocols. We will illustrate the process by using an AE Pro equipped Apple to call another one so equipped. (But the basic routine would be the same for any two computers running software supporting this protocol.)

First, connection is established. (If two AE Pros are involved, the one being called is simply left in the unattended mode via the " + " command. The other dials with the "D" Main Menu Command.) The calling computer supplies the password at the "Entry:" prompt and at the ") > " prompt the transfer is ready to begin.

We will say that we are transferring a program written in Applesoft called MORTGAGE.FP via these protocols. The first thing to do at the remote command prompt is to use the "D" remote command to see the directory of the disk running on the called computer. Reading the disk catalog would show that names were already in use — we'll assume we can call the transferred version MORTGAGE.FP also.

Now, still at the remote command prompt, type R for Receive, and the called computer will get ready to receive a file under the protocol standards. The called computer will prompt for a name to call the file:

filename:**MORTGAGE.FP**

As shown, we answer with the name of our file as we want it to be stored on the called computer's disk. Now we have to send or upload the program from our calling computer. In the case of AE Pro a control-Q will get us AE Pro's on-line menu prompt (+ >) at which we enter S for Send. AE Pro then prompts:

[S]tandard or [P]rotocol?

Instead of answering with an S as we did when uploading a file to Compuserve's FILGE editor in the last chapter, we now pick the P for Protocol option — and answer the next prompt for filename with the name MORTGAGE.FP (or whatever name the program was stored under on the calling disk.)

The two computers then begin to pass the information from one to the other. AE Pro will, on the calling computer's screen, print the number of each block of copy being sent, tracking the transfer's progress. At the conclusion of transfer a CRC is done automatically on both the original and the transferred copy and if the resulting numbers from these two CRC operations match then the transfer is considered a success.

But suppose that the terminal used on either end does not support protocol transmissions? How then can you upload and download program files from one to the other in addition to ASCII text files? The answer is that, in most cases, you would not be able to without first changing the program into a transmittable form. In order to send, for example, a BASIC program from one computer to another without using protocols it would first be necessary to change the BASIC program into an ASCII text file. Then, once the ASCII file is sent, it must be reconverted from a text file back to a program.

Most terminal programs have utilities that enable you to perform these extra steps. Without such utilities, you would first have to manually convert the program to an ASCII file. To do this on our system, add lines to the beginning of the program which will open a file on your disk and list the program to the new file. Consult your Apple DOS manual. At the other end you will have to move the ASCII file into memory—which you can do in the Apple environment by using the EXEC command from the Disk Operating System. As you can imagine, each of these steps opens you up to one more realm of possibility for making an error—and such an error may not be detected until long after the transfer has been completed.

There is no doubt that for transferring files of any sort, it is far better to use the protocol method if it is available.

Most terminal programs allow you to send and receive text files and programs. Some terminal programs even support very sophisticated auto-transfers and error-checking capabilities.

But suppose you want to communicate with more than just one other computer and you want all the other calling computers to communicate as well? And, in most businesses, transferring programs is unimportant. Pertinent information retrieval involves such things as reports and inventories, directories and bills, all the information that can be stored in a computer file.

PRIVATE BOARD

If your business or other use demands that many people have access to your computer system's files and programs, and if the callers should be able to leave messages for each other, and if you need a level of security (perhaps even different levels for different callers) then you need a private bulletin-board and remote-controllable system. An example of this might involve a number of salesmen in the field. Each salesman might be equipped with a portable terminal or even a computer and modem. You would want all of your sales force to be able to access files showing the latest inventory of the main warehouse—as well as to be able to read the latest message files from the main office. The district managers might want to be able to send private messages to various salesmen. The regional managers might want the capability of updating some of the files that the other people will be accessing. And, finally, the central office would want complete—and secure—control over all of the above.

Tasks like this, thanks to the development of new software packages which have taken the art of telecommunications a conceptual level forward, can now be accomplished with astonishing ease. For our do-it-yourself system, we have found the Online program (from Southwestern Data Systems) to offer many advanced features for those who are looking to set up a password-protected, private bulletin-board and remote controllable system featuring various levels of security. The Online program will allow approved account-holders to call the main Apple system. Moreover, they do not have to be equipped with Apples, but may have any computer and modem combination or dumb terminal.

Once connected to an Online system, members with low security (security level 0) can send mail to other accounts (if they know the other accounts' names) and read mail only from the system operator (sysop). The next level of assigned security (Level I) will allow access to board-public text files. The next level of security (Level 2) may access any file, not just those intended for Level 1. The top security-clearance is Level 3 and from this level not only may text files be accessed—any program stored on the main computer's disk drives may be remotely loaded and run. All users, sending messages stored as text files, have access to a small but relatively powerful word-processing text editor.

The Online system will support a minimal system of 48K RAM, (either an Apple II Plus or an Apple II with Applesoft on card), one disk drive and the Hayes Microdem II. But, the addition of a second disk drive will prove almost mandatory if you intend to use the system to run programs remotely. And, a clock board is also recommended (ones supported as of this writing are the Mountain Computer Clock Card and the California Computer Systems Clock Card).

To see how this system works, we'll start with the lowest level of security and work our way up to the top. Keep in mind that higher levels of security have available all of the options of each lower level.

If you are not using our do-it-yourself system, following this capsule summary of the Online program should serve as somewhat of a buyer's guide. Check what software is available for the computer you are using and try to get as many of the same style of features. Although, as of this writing, the Online system was one of the most complete security-capable system for any computer—it is likely that new software will be developed for the other popular computers as more and more people become interested in using telecommunications.

Level 0 is considered a guest level. Ordinarily the sysop, or other level 3 personnel, will set up an account name of "Guest" and a special password. So, when the caller is asked by the Online program for his account name, he replies with "Guest" and the password supplied to him. (Obviously, Online will not allow access by curious callers who may have found the number but are lacking the password.) Although Level 0 is able to use many of the commands of the Online system, this low security level only allows the guest to receive messages posted (usually explaining guest privileges) from the sysop to any guests and to send messages to an account name. Level 0 cannot access any of the text files that may be stored on the disks. But, Level 0 at least allows a person to become familiar with the Online main commands.

Here we will only examine a few of the most powerful of the commands available and briefly summarize the others. At this point, it is more important to comprehend—in a broad sense—what this Online package allows telecommunicators to do.

The first thing a caller to an Online system will want to do is to check and read his or her mail (A level 0 guest will only receive mail from the sysop). From the mail functions, a caller may also want to send mail. The M command checks mail waiting for the caller's account name (keep in mind that all guests can have the account name of "Guest") and allows retrieval of that mail. At the end of each piece of electronic mail there will be a disposition prompt:

Disposition (a,d,r,?)

At this disposition prompt you may enter A for read the piece of mail Again; D for Delete it; R for Reply to it; or a ? to display what the options are.

Level 1 is the security level that will be most often used by callers. A Level 1 caller is a person who is known to the system—a person who you expect will be calling, possibly a member of the business firm or organization using this Online bulletin-board. The level 1 caller has all of the same capabilities of Level 0 and more.

Like a Level 0 caller, he or she can receive and send mail. But, mail may be received from any other account name. Like a Level 0, a Level 1 account name has access to all of the most basic system commands and can set such things as linefeeds, lower case, display of time and date, brief mode, and such.

Level 1 is the first level which has access to the sytem's line editor. It is also the first security level which is able to access and read text files stored on the disk drives.

The line editor is one of Online's most important features. It is similar, but not identical, to the line editor within our system's AE Pro terminal software. Like that editor, it is a line-based word processor. To correct a letter or text file being entered you list the lines out to the display and each line is prefaced with a number for reference. By referring, with various commands, to a line of text by its number you can change, delete, or add and insert lines of text as you go along. Editing a line of text may involve retyping a word or two using the change commands, or adding words to the beginning or end of a line.

Particularly useful is the line editor's U for Upload command. In this mode, the Online system will accept into the editor a file which is being sent by another computer. It does not matter if the other computer is running AE Pro, or even if it is an Apple. The Online editor's U command will accept any ASCII file which is being sent at the usual 300 Baud rate. So, a Level 1 user with a TRS-80 system or with an IBM PC or with an Atari 800, with a DEC Rainbow 100—with any sort of computer and modem—can share informational throughput with this unique and hospitable system.

With the U command it is possible for the most distant users of the Online system to avoid long connections at expensive long-distance rates. The caller first composes his message offline using a word processing program of his choice on his own computer. The only stipulation is that the file is an ASCII encoded text file. Then, when he calls the Online system — the Online system will accept his upload.

The editor will also work as a transmitter as well as a receiver. It can download its information to any calling computer. The .LN command will send the file which is in the editor to any calling computer. The other computer would then receive the file as an ASCII text file. (There is even a method to send a program directly into the calling computer's RAM memory—if that calling computer is an Apple with a Hayes Micromodem.)

Another feature of this friendly editor is that it allows for the vagaries of phone connections and power companies. Let's say you are uploading a long file. In the middle of it, the phone company suddenly cuts you off. No problem, if you call back immediately the Online system will have saved your data up to the point the connection was broken. And, if the Online system itself loses AC power, the computer it is running on will of course turn off. But when power is restored, the Online system will automatically come back to life.

Once typing or uploading the message into the editor is completed, there are a number of alternatives. You may go back into the mail program and send the message to another account name—or to a number of account names. Or, you can put the message to disk as a text file. Once it is there as a text file saved at a Level 1 security level, it can be read at any time by anyone who checks into the system with a greater than 0 security level.

This feature of accessibility to text files stored on disk is one of Online's most interesting and easily-tailored abilities. Two commands allow the caller with Level 1 or greater security clearance to read text files that have been stored to either of two disk drives (the !R reads files on the first drive and the !!R command accesses the second drive.). By using the editor to leave text files, and these commands to read them, the Online system allows all members of the board to communicate important and/or timely information to all other members. Coupled with a feature which fully supports private mail delivery between account names, it becomes reasonable to say that the Online system can become any corporation's first electronic switchboard to keep remote locations in direct contact with a central office facility.

Level 1 members, however, may not ordinarily know all the files which are available on the system's disk drives. It is up to the sysop to inform those who need to know the names of files that they will be able to access. One way to do this is, as the manual details, to set up a special system file. Title the file, for example, TITLES. Then, when a calling Level 1 or greater member types TITLES at the main command prompt he will see what files he has available for access. Note that there may be many more files on the disks than the sysop releases to public view.

A Level 2 member not only has the ability to access the public files, he can also see the complete catalog (list of files) on both of the system's disks. And, he can access all but the files that are used by the Online system. This obviously powerful position should only be granted to members who should be able to have unlimited access to any information stored on the system.

Level 3 security clearance, the top level, is ordinarily granted only to relatively few people. This is because the top level not only gives access to all the files on the disks, it also gives the caller access to the Online system files and will allow the Level 3 caller to run any programs on the disks just as if he was seated in front of the main Apple's keyboard.

This last feature—the ability to run programs remotely—is a real wild card. If your business or organization depends on special cost formulations, invoice calculations, or complicated projections—all those functions normally performed by the computer in the office—much of this can now be done from anywhere in the world through a call to the Online system.

Level 3 includes special commands which allow the caller to exit from the Online main or editor programs directly into the Basic language of the host Apple computer—all from a remote location. And, again, it does not matter if the caller is running an Apple—he will still be able to control the host Apple.

There are, of course, certain rules that must be followed as to the type of programs that can be run remotely. The most important limitations are that neither the Apple's graphics nor its special display commands can be sent over the phone line, and the program to be run remotely cannot do anything which would switch off the Apple's modem capability.

The Apple's low-resolution and high-resolution graphics are dependent on the Apple's specific setup of display memory. As the calling computer can be anything from an Apple to a dumb terminal with a paper printer no graphics other than the text-style graphics can be supported. (But, for some information on future possibilities of graphic transmission including colorful Telidon graphics and realistic slow-scan television see Chapter 12.)

Display commands such as VTAB and HTAB which move the Apple's screen cursor about also cannot be used here. And, again, this is due not only to how these commands work but as to standardization between remote and host computers.

Finally, commands that access the I/O ports of the Apple (the card-slots that you plug circuit boards into on the Apple's main circuit board) cannot be used within programs that are to be run remotely. The reason is that, when the Apple is set up as a telecommunicating device, the modem in port #2 is in use. If you switch to a different port—you've hung up the modem! The main thing to be sure of is that none of the programs on the system than can be accessed and run from a remote location use any other ports in the Apple. This means that any program that uses a printer is excluded. If you try to access a printer remotely, you will lose contact with the Online system.

The Online system fulfills just about every need as to security and access that a corporation or organization might want. But what about when security is less a factor than a wide-ranging ability to share information?

PUBLIC BOARDS

The answer to your organization's telecommunicating needs might just be to set up one of the Public Access Message Systems (PAMS) dedicated to your own needs. These are the systems that we detail in Appendix One, and which include the PMS system which was examined in Chapter 3 ("Your First Phone Call").

Using a PAMS for your organization will have some benefits and some drawbacks—these systems are specifically designed to be open to anyone who calls in. On the plus side—this means the audience is very wide. On the negative side—security is almost impossible to implement.

Let's assume that you are running a mail-order store. Obviously, the more people you can get your catalog in front of, the better. In this case, a PAMS should be very effective. You could set up the store or mail-order catalog as one of the files that anyone can read. And, you could use the message area to exchange information with potential customers. Of course, there will be a lot of personal messages on too—callers talking to each other. But, the goal of getting a catalog before the public will have been met. There is even a modicum of security provided in most of these various PAMS in that a message may ordinarily be password-protected to be private between the sender and recipient.

But an investment firm would be foolish to use a PAMS-style electronic bulletin-board. Files which would best be kept secret within the corporation could be read by anyone. The constant chatter on the board would fragment and make meaningless any serious attempt at business conversation. This type of firm would be far better off with a system such as Online.

Many of the PAMS systems, such as the PMS and others, do allow uploading and downloading files with great ease, as well as setting up long files for all to read. For an organization which would not mind trading security and a business-like demeanor for access to a large audience of computerists, a public board should be investigated.

Read over the help files in Appendix One. Note which of the various board styles seem best to suit your purpose. Then, call a few of the boards that use the style of software you want to investigate. See how they work in a real-life situation.

11. Some Advanced Techniques

The methods outlined previously should stand any telecommunicator in good stead. You are now able to get your message across—or to receive messages and information from others—either on the national computer networks or by using one host computer to hold together a private communications net of your own.

But, there are many ways to make this process easier and more reliable. Some of these methods will become obvious as you work—and live—in this new society of cheap, reliable, and virtually instant computer communications.

DATABASE YOUR TEXT FILES

Certainly by this time you have received and used quite a bit of information via your computer. And, the odds are, one item at least has already been "lost" and (knowing that Murphy's Law most definitely applies to computers) probably this loss has caused some inconvenience. The situation can only get worse. As it becomes routine to log-on to the information utilities with a fresh disk in the drive and a capture buffer running wide open, disk after disk full of text files will be generated for storage and later perusal. And, as a private PAMS or business board attracts or is assigned more and more members—an ever-larger quantity of electronic mail and computer-generated messages will be received. Escape from an avalanche of paper often leads to being buried in a collection of disks.

The partial solution for this is to find a "universal" database which can be interfaced to the system's terminal software. Such databases are available for all personal computers. The computers most likely to be used with a modem happily have the most number of databases to choose from. The Apple II, Atari, Radio Shack, IBM, Commodore and other popular machines have attracted a good deal of secondary software support.

Databases come in all price ranges and with all kinds of features and special options. Some of these features and options are absolute "must-haves" for the telecommunicator.

First, this is what we mean by "universal." Many ads appear in the computer magazines for databases that keep shopping lists, can inventory a small wine cellar, or can bring order to a chaotic record collection. These type of databases do only one thing—they are called "dedicated" databases. They are only capable of one job, though they do that exceedingly well (usually). These kinds of databases are not relevant to telecommunications because of the fact that the vast amount of information and messages—all that can be downloaded and stored as text files—will cover far more than one area of interest or applicability.

Instead, a universal database allows the user to design various formats for separate files. One database format can be designed to keep track of magazine contents pages, another format will keep track of a billing list, yet another format may store a phone book—and so on. Each time a new format is made, the user decides what each separate database file will contain.

The user first must decide upon the main categories of information. These very general categories become separate databases. For instance, a professional might want to keep the addresses and phone numbers of his personal friends in a separate database from the business records of his clientele.

Once it is decided what will go into one of these database files, then the ways of structuring each entry is the next step. An address-book file, for example, might contain names, street addresses, cities, zip codes, area codes, and phone numbers. For each person, all the variables together with the person's name would be one record within the database. The various parts of each record (such as the phone number) are called *entries*.

Some databases are very formal. They require the allocation of exact lengths for every entry in a record within the file. When an entry is defined as being a person's name it is necessary to (one-time only) decide what the length of the longest entry might be. The length of an entry is referred to as that entry's field length. A typical field length assignment for an entry of Name is 25 characters. Such a field for a names-style entry usually allows for just about every name (though some first names may have to be abbreviated).

A formal database, because it is so structured, allows for the user to utilize every possible byte of space on a disk. And, these databases usually allow for searching and sorting records easily.

But the formal databases are not the only style of universal database. Nor are they, we feel, the best way for most people to go about setting up databases with which to collect the eclectic mass of data the information utilities will provide.

Many universal databases use an "informal approach." In an informal database, the first choice left up to the user is still the same as in a formal database structure. Separate database file titles have to be thought of which all divide the information flow into logical and useful categories. Once these main file structures are decided upon things are quite different in one of these informal data bases.

In an informal database, one record may have a style of entries wildly different from another record in the same file. And, there are no very strict field-length rules. Each recohd is set up more or less as it is entered. In an address book file, the order of the entries of street address and phone number could be switched about and some records could contain additional entries. In one of these informal databases, a record can simply be a screenful of information.

The informal databases are usually easier to learn than the formal databases. But, they are more wasteful of disk space than the formally structured ones and are often not quite as powerful. Still, for telecommunications work, we do recommend the informal style (though we will also, below, be telling you how to apply both formal and informal styles).

For most people, the modem and computer usually open up realms that have never before been investigated. It is surely difficult to design a structured format to accept information which has not yet been generated, and the format of which is unknown. And, while some information being downloaded is neatly formatted, with every record having the same number of entries and with maximum field lengths intact, other information—particularly the interactive communications—has no real database-style structure.

Electronic mail is usually formatted in paragraphs much the same way that a letter on paper is communicated. A formal database would be unable to accept and use this type of information. An informal database, on the other hand, would simply treat the letter as one more full-screen record within the file.

Choosing a database to use with a terminal program can be a most difficult decision. First, the decision must be made as to whether formal or informal is appropriate. Then it must be made absolutely certain that the database can accept information in the form of ASCII text files. It should never be necessary to have to retype information that has been downloaded. Instead, the downloaded text should be automatically accepted into the database. The database must share the same disk operating system with the terminal program—otherwise text files in one program will be inaccessible to the other.

Most users of the national computer networks and the hundreds of Public Access Message Systems find that almost all of their downloaded files are quite different from one another. One day's worth of downloaded files could include recipes from a cooking SIG, electronic mail from your corporate branch office and the latest baseball scores—but the next day's files will include everything from the Standard and Poor's summaries of your portfolio's heaviest ten companies to amortizing the mortgage on a client's home. Obviously, the formats of all these various forms of information will have no common thread. Using a formal database would require spending a relatively large amount of computing time designating formats to hold all the information. This is why, for many people, an informal database is certainly their best bet.

Let's take a look at the *Visidex* database from Visicorp. This is an informal database for the Apple and the IBM PC (it may be available for other computers by the time this book is on the shelves). Visidex fulfills almost all of the functions needed in a database which is capable of directly accepting text files that were originally downloaded onto a disk by a terminal program. (Visidex will, of course, work hand-in-hand with text files downloaded by AE Pro.)

Visidex is a very powerful program whose accompanying, thick manual explains its operation every step of the way. Here, we will only be concerned with the most general description of this database—just enough so that you should be able to tell if it will, indeed, solve your own problems about what to do with important data.

Visidex organizes itself by "screens" of data. Each screen is an informal record. From the keyboard, it is possible and easy to type just about any format onto a screen. Information is typed just as if it were being entered onto an index card—and thinking of Visidex's screens as video index cards is one way of recognizing their potential for storing all sorts of informational formats.

While in the Edit (typing-in) Mode, Visidex offers a number of one-key commands to delete lines once entered, to save lines in memory to insert again later, to use normal white-on-black letters or to make an interesting video display utilizing inverse or flashing characters. At any time during entry, other commands move an entry-cursor to a different area of the display so that the visual format can be made very fluid. All commands are referenced at the top of the screen and so are usually very easy to memorize after a few sessions.

Once the screen is designed so as to have the information presented in the required format—the next step is to choose which of the words on the screen will become the screen's *keywords*. The concept of keywords is what Visidex bases its powerful features upon.

Screens may be searched through quickly by keywords. Entering the keyword "recipe" during a search would—at the end of the search—retrieve each screen which was formatted with the word "recipe" as a keyword. Searching may also be done on combinations of keywords. A Visidex database file could be searched for screen records which contained the combination of keywords "recipe" and "appetizer." In this case, the end of the search would display for the user recipes for appetizers, but not those recipes whose screens did not have the keyword of "appetizer." A screen may have many keywords. Indeed, the more keywords that have been called-out on each screen then the more powerful Visidex becomes as searches begin to relate items via keywords between which—at first glance—the user himself would not have realized the connection.

Here is how to use Visidex in connection with ASCII text files that have been downloaded via a terminal program:

Visidex will read-in text files from a disk by means of a special control-R command. When in Visidex Edit Mode, the keyboard entry of a control-R begins this process. First, a message will explain that two disks are needed (one with the text file to be read, the other a Visidex data disk) and then another message will ask for the name of the text file and which drive it is on.

When this prompt has been answered and Visidex has found the text file, each successive time that a control-R is pressed at the keyboard another line of text (from beginning of the line to a carriage return) will be read from the text file into Visidex and appear in the screen record that is being edited and viewed.

As control-R is continually pressed to read in the file, the end of the file will finally be reached. There will be an "END OF DATA" message. Stop pressing control-R. But, if another control-R is inadvertently pressed, Visidex will appear to "crash." There will be an error-message generated at the top of the screen saying "DISK I/O ERROR" and another message at the bottom of the screen reading "HIT RETURN TO CONTINUE." Do *not* hit return! If return is pressed, Visidex will reboot the data disk, blank out the present screen and lose all of the information which has thus far been read from disk.

Instead of hitting return—hit the escape key. This will—as always—accept the present screen and then drop back into the Editing mode. From the editing mode it will be possible to clean up the screen, then go to keyword mode to assign words on-screen as record keywords, and finally to be able to save the screen record to disk.

More than one reviewer of Visidex has stated in the computer magazines that, because of the seeming crash of the program, they could not use Visidex to read text files. But, as my own Visidex files attest to—it is both possible and simple. Just do *not* hit return if a mistaken control-R gives you an error at the end of file—hit the escape key instead. Or, as an addendum to the manual states, it is even better just to know what the last line of a file will say and not to control-R past it in the first place.

Suppose that a friend has been requested to send EMAIL with his name and address for your files. You log onto Compuserve the next day and, sure enough, he has reponded with an EMAIL. The menus or GO-EMA-3 lead to the EMAIL as, with capture buffer on, a message similar to this below is received:

Hi There,

My address is 1234 Sycamore Blvd.,
Megatown, NY 11714. By
the way — I didn't see you on the
CB Simulator last night. Be there
tomorrow?

Best,
John Doe (Computerman 5)

Downloading the captured message to disk would result in a text file which Visidex would be able to access. But there would be some drawbacks to using this just as it is. The most obvious would be that all the extraneous information would take up space on the data disk best left for more important information.

Now, in Visidex, that database's control-R command may be used to read in a complete text file. After a number of control-R's the screen record is now the same format as the received EMAIL. But all the information need not—and should not —be saved in the database. Only the parts which may have to be referenced later make any sense to save.

Visidex's many editing commands could be used to delete lines and move other lines about so that the final screen record would read simply:

John Doe
1235 Sycamore Blvd.
Megatown, NY 11714

At this point, keywords such as "John," "Doe" and perhaps "11714" could be defined. This, if such keywords were defined throughout the database file, would allow for searches on first and last names as well as on zip codes. By deleting and reformatting, the screen record has not only been made more compact for more efficient disk storage—it also been made easier to read and reference.

Visidex will not be the solution for all database and telecommunicatng problems. It has may limitations. The most important limit is the amount of information which may be stored as one screen record. Each Visidex screen is 40-columns across by 20 lines deep. Using the editor subroutines from within a terminal program, or accessing files with a complete word-processing program, may be necessary before a file can be loaded into Visidex.

And Visidex does not offer the arithmetic power of some of the formal databases. It will not, for instance, print out long, columnar reports and mathematically process one field by another one. Visidex is more a database for words than it is one for numbers.

If a relatively large portion of downloading time is spent within one area, and if that area is a numerically-based datasource, a formal database may make more sense than an informal one. The main thing to remember is not to get involved in a situation which would require many formats to be defined.

If a database can be defined for the records which will be downloaded as to field lengths, visual format, titles and such, then it is possible to use a formal database, although it will be more complex than an informal structure. As a formal database is complex and changes from one use to another, we can give only the most general hints about its use.

Most formal databases do not have the capability of reading information from previously processed ASCII text files. They may store their own data in a non-ASCII manner, they may use their own disk operating systems, or they might not be able to accept input from anything but the keyboard. For these reasons (and any combination of them) it may be difficult to locate a formal database which will easily interface to a terminal program.

For our do-it-yourself system we have located one such—*The Data Reporter* from Synergistic Software. Although this program is copy-protected it is uniquely user-modifiable. The BASIC portions of this program may be listed and modified, and then saved to the disk.

One of the modifications which are listed in Section 9.2 of the users manual, "Selected Program Modifications," specifically details the changes to the Data Reporter program that must be made to read in text files from another disk. These changes involve typing in a few short lines; but great care must be taken that they are entered exactly as shown in the manual.

Databases are one of the main ways that a personal computer can organize information, detail by detail. As more and more information is received by computer from other computers—it will become ever more desirable to have a database program with a telecommunications capability.

BEATING THE CLOCK

Spending time is often the same as spending money in telecommunications—as we have often said. As Appendix One lists, the hundreds of Public Access Message systems throughout the country are an enticing batch of potential phone calls. Sooner or later a favorite system will turn up that not only isn't in the next county, it may not even be in the next state. And, though the computer networks are extensive in reach, many people do live further than a local call away from the nearest information utility port.

But there are ways for the computerist to lower long distance telephone costs by taking advantage of the capabilities of his personal computer. Many computers either have built-in timing functions or can accept optional "clock cards." By using the clocking facilities together with proper software, it is possible to accept information from distant computers even while you sleep.

First, a program must be written which will continually "poll" the computer's clocking facility. Depending on how far in the future the call will be made, the timing mechanism will at least have to know the hour of the day, perhaps even the day of the month. This internal clocking scheme must not interfere with the operation of the computer's modem.

Once the proper time has been reached, and this fact is recognized by a "caretaker" program—the caretaker program must then trigger the phone call. At this point, you may be able to interface your caretaker program to an existing terminal program which will then place the call for you.

The do-it-yourself system nicely illustrates the steps along the way to this kind of automatic communication. The Mountain Computer Clock Card was originally mentioned in the previous chapter as an optional accessory for the Online software package. Other clock cards are just as feasible to use — but the Mountain Computer card does seem to be the most widely accepted among the secondary software suppliers.

The card simply plugs into any of the Apple's I/O slots (other than slot 0). The manual details how to set the time using switches on the card in conjunction with supplied software. While it is plugged into the Apple, the card is always charging an on-board nickel-cadmium battery. This is helpful if the computer is not going to be left on at all times as the battery-backup will keep the clock running between sessions of use. Without such backup, and if the computer were shut down each day as is often the case in most homes, quite a bit of time would have to be spent resetting the clock.

Once the clock is installed and set to the right time we are ready to begin designing the small caretaker program. All that this program will do, in its simplest state, is to keep looking at the clock.

Most clock cards and built-in clocks will allow even a BASIC language program to take full advantage of their features. The Mountain Computer Clock, for instance, puts the time into a string variable called T$. The T$ variable contains the information in the following format: MONTH/DAY   HOUR;MINUTE:SECOND.FRACTION. A typical reading of this T$ variable would show the program reading the clock that T$ = 01/14   10;15;13.456 which would indicate that it is January 14 at 10:15 AM plus 13.456 seconds.

At this point, if there is no terminal program with automatic calling capability available for a caretaker program to trigger it would be necessary to write a longer and more complex program. This program would have to dial a number, connect, recognize when it was being prompted by the host computer for a command, send the proper commands, get the information, store it to disk and perhaps print it out, then log off and hang up the phone. With some computer systems, this procedure simply may not be possible. The type of system which uses an acoustic modem, where the phone connection is made by a person physically dialing a telephone, is obviously not at all suited for this. Only systems which have modems that are direct-connect to the phone line and which can be dialed under program control can be used for this method.

In our recommended system, most of the work has already been done. The Hayes Microdem II direct-connects to the phone line — and the AE Pro terminal package is capable of performing all of the automatic routines that will be needed.

AE Pro may be configured to automatically dial a number. Then, use the conditional macro capabilities to send commands at the right prompts. Remember that AE Pro will "handshake" on incoming words. In this handshaking mode, it will not send a command until it has received the proper prompt. AE Pro may be configured with the Install program to come up with the printer on so as to keep a hard copy of information. Or, you can have it autosave information to disk as the capture buffer overflows.

Always keep in mind, however, that without someone watching the connection — there is always a good possibility for failure. There are many variables, all of which must be working together perfectly to succeed. If, for instance, the exact command structure of the called computer is not known—it does not pay to guess.

Without knowing to the very letter what every single prompt encountered will be, and what exact command should be given to each of these incoming prompts it would be foolish to expect the system to work. And, never try to download sensitive information in this manner unless it will be available for a second try.

But, if the logical and orderly programming procedures are followed for interfacing a computer and modem to a clock, the final result could be a deeply slashed phone bill.

KEEP BUILDING-BLOCK FILES

Earlier we explained how a terminal program's editor or a word processing program could be used to compose letters and documents while off-line to be transmitted later over the computer. As long as the documents were stored as ASCII text files, they could be sent via modem. But sooner or later the time comes when it is just not convenient to log off, compose your message, and log on again. Perhaps the phone number has been busy and there is doubt if a second phone call would get through.

A solution for this is to give a little thought to the types of messages most often sent. Many people discover that they are sending twenty or so "standard" messages — and that even these messages contain certain standard parts.

For instance, you might always leave the same style of message when you check into a PAMS for the first time. Certainly, your name and address is always the same, as is your phone number and other personal information.

All of these standard parts to various messages should be stored as short text files on the same disk with the terminal software. When it is necessary to send them or add them to a message they can be quickly uploaded directly from disk. While this will not be as cheap as if the message had been composed while entirely offline, it will be far less expensive than to have taken the time to type in the same amount of information.

ADDING VIDEO PICTURES

Thus far, all telecommunications we have seen have been through words and numbers scrolling across the screen. While the utility and usefulness of such communications is without question, we are still very visually oriented creatures.

Many people find that using a computer and a modem will allow almost all of their business to be conducted from one remote office to another. But there may often be times when it is suddenly very necessary to see a chart or a diagram, or even a picture of some component. At that point, it may seem that a barrier has been reached and that the art of telecommunications may indeed be a limited one.

But, such is not the case. It is both feasible and simple to add pictures to computerized communications. It is even relatively inexpensive. There are two methods of doing this, each more suitable for some things, less suitable for others. You may need both, just one, or you may be perfectly satisfied with the already amazing capabilities contained within pictureless video.

The two methods are slow-scan television (SSTV) and Telidon videotex. Both of these methods allow for graphics to be sent via the same telephone connection over which the computers are exchanging information.

SSTV is a means of sending black-and-white television "frames" over the phone. It is unlike broadcast television in that each full-screen frame takes eight seconds or longer to be received. The length of time a frame takes to be sent is related to the resolution—the clarity—of the image which the equipment used is capable of delivering. Pictures in motion cannot presently be sent through a phone connection as they require far more bandwidth (or signal "room") than do these slow-scan pictures.

The Robot Corporation makes a number of well-engineered SSTV terminals. One, the Robot 400, is manufactured mostly for radio amateurs but it may be easily adapted to a business use. Other terminals from Robot are more expensive but offer pictures that are of finer detail.

The 400 has a resolution of 128 by 128 pixels. A pixel is a picture element; each of these rectangular picture elements may have one of 16 shades of gray. A broadcast TV image has far greater resolution than this, and offers almost unlimited gray scale.

A video camera (a cheap black-and-white, non-sound one is all that is needed) is simply attached to the back of the 400. The operator adjusts two controls on the front of the 400 to achieve the best picture on-screen. Then, this on-screen picture is automatically converted into SSTV graphics. The operator at the other end of the connection, on his own SSTV terminal, receives the image. Because of the large pixels, the image will be blocky-looking. But we have found that type sizes of 12-point or larger can be read, expressions can be easily determined on faces, and a surprising amount of detail can be counted on.

The 400 requires a phone patch (available from your phone company) to attach to the telephone lines. This connecting procedure is somewhat complex and should not be attempted by anyone unfamiliar with such things. If your business does not have access to a person with this capability—the Robot Corporation also makes a number of terminals that connect directly to the phone line. These terminals, more expensive than the 400, also offer better resolution.

We found the 400 and its associated models to be an excellent means of adding black-and-white video from a camera to a telecommunications setup.

But what to do if an application demands bright, colorful and eye-catching graphics? The solution to this problem lies in a new method of videotex based on the Telidon standards.

Videotex allows for the sending and receiving of full-color graphics—drawings, charts and pictures—over a normal phone line. The problem in the past was that there have been numerous non-compatible means of doing this. The English Prestel standards, the French standards, Canadian standards, experimental methods all have contributed in the past few years to make the world of videotex graphics seem a most unsure one.

Recently, AT&T studied the situation and released what it intended to be the standard for such telecommunications activity. They chose as the new American standard a method called PL/P. The PL/P standard does include all of the Canadian Telidon protocols. So, a Telidon system will—in all likelihood—be able to communicate directly with many new computer networks in the future as well as with other Telidon computers and terminals today.

Norpak, a Canadian firm which has had much experience supplying stand-alone Telidon terminals and computers, has entered into a new venture with

Apple Computer, Inc. to produce a plug-in board for the Apple which will give our system almost complete Telidon capability. Briefly, the Norpak Telidon card enables an Apple-based system to send and receive high-resolution graphics over the phone. Instead of the Apple's normal six high resolution colors with a resolution of 280 by 192 pixels, the Telidon system delivers a full 15 colors at a reduced resolution of 128 by 96 or a black-and-white image at a resolution of 256 by 192 pixels.

Even a non-artist can achieve excellent effects with this new system. The software with the system allows the video graphics to be generated with a simple joystick after a few commands and command structures are learned. The Apple takes over such things as repeating shapes, combining forms, even drawing smooth arcs and straight lines between points. If you can picture in your mind what you want to see on the screen, the odds are that you can achieve it—even if you, literally, cannot draw a straight line with a ruler. Although the resolution is less than a typical (and far more expensive) Telidon terminal, it is absolutely acceptable for Apple-to-Apple communications. It even delivered good results when communicating to a large Telidon database network.

The Slow-Scan TV method can allow you to send the type of black-and-white video effects you can achieve with a simple video camera. The Telidon system will let you design full-color screens of pictures and diagrams with text included. The methods are by no means interchangeable, but rather each offers capabilities to support the other.

While the future of telecommunications may include everything from three-dimensional imagery to synthesized voice-programming—even today's possibilities are immense. Words, data, and pictures can be sent via the phone line and there is no doubt at all that those who can use this impressive new art will have the advantage.

The advantage is yours.

APPENDIX

Appendix

Public Access
Message Systems

In the centuries when printing first began, the precursor to today's newspaper was the "broadside." Published, for the most part, in people's homes or in small printing shops these leaflets carried every kind of thinking—from revolutionary maxims to religious canons—into the hands of the people.

Today's answer to the historical broadside is the computerized bulletin-board service often referred to as Public Access Message Systems (PAMS). Most of the PAMS listed here run on computers in people's homes or in small computer stores. Some are filled with messages of purely personal nature, others carry much in the way of programming information, some are political—they are the leading edge of the new information explosion. As such, they are interesting both as phenomenon and as a source of unique telecommunicating.

Just as some broadsides were printed in various type fonts on assorted styles of paper and in many formats, today's PAMS all run on various computer systems. But, just as any broadside could be read by anyone with eyes and the skill of reading, all of these PAMS may be accessed by anyone with a computer (or terminal) and a modem.

But, each style of PAMS has its own command structure to learn. Most of them share certain commands ("H" or "?" on most will get you a help file to read), but some are quite different in their structures. Because you call these systems directly, you are often in the costly position of learning how to "read" them while paying a long-distance toll. To help out, we have downloaded some of the help files from the major styles of PAMS and have printed them following the phone number lists.

The first phone number list details the country's active PAMS grouped by type. The type usually involves what software the PAMS is using and what kind of computer you will be calling. Apple systems such as NET-WORKS, ABBS and PMS seem to attract more in the way of Apple-using callers while systems such as BULLET-80 running on TRS-80 computers attract more in the way of TRS-80 users. But in the final analysis, most of these systems have every style of computer-user calling them.

Your best bet is to hunt around first in your own area code and then gradually move further afield to find boards that have content that will be of interest to you. To help you do this, we have sorted the phone number list on area codes. You will find this list following the first.

Like all lists, this one will gradually become dated, although many of the boards have already been operated for years and most are established on a permanent basis. Up-to-date lists (by style only) can be downloaded from Bill Blue's PMS system as noted at the beginning of the first list.

For now — here are the PAMS systems that are available for you to call.

PUBLIC ACCESS MESSAGE (and file transfer) SYSTEMS
(P.A.M.S.)

Compliments of People's Message Systems, Santee CA.
(714) 561-7277

        *24 denotes 24-hour operation
        #1 denotes original system of that type
        -rb denotes call, let ring once and call back
        -so sexually oriented messages
        -rl religious orientation
         ! new system or new number to existing system
         $ Supports VADIC 1200 baud operation
         & Supports 212A 1200 baud operation
         % Supports BAUDOT operation

Please send updates/corrections to:

PMS Santee, TCB117, 70315,1305 or BBLUE

And to:

The Small Computer Connection
c/o McGraw Hill Book Company
Technical & Reference Book Division
1221 Avenue of the Americas
New York, NY 10020
ATTN Neil L. Shapiro

LISTING ONE

ABBS	ABACUS II, Toledo, OH	(419) 865 1594
ABBS	ABSS, Dallas, TX	(214) 661 2969 *24
ABBS	AGS, Atlanta, GA	(404) 733 3461 *24
ABBS	AIMS, Oak Brook IL	(312) 789 0499 *24
ABBS	Akron Digital Group, Akron, OH	(216) 745 7855 *24
ABBS	Apple Cider, Las Vegas, NV	(702) 454 3417
ABBS	Apple Crate I, Seattle, WA	(206) 935 9119
ABBS	Apple Crate II, Seattle, WA	(206) 244 5438
ABBS	Apple Group N.J., Piscataway, NJ	(201) 968 1074
ABBS	Apple-Med, Iowa City, IA	(319) 353 6528
ABBS	Apple-Mate, New York, NY	(201) 864 5345

ABBS   Baileys Computer Store, Augusta, GA . . . . (404) 790 8614
ABBS   Baton Rouge, LA . . . . (504) 291 1360
ABBS   Byte Shop, Ft. Lauderdale, FL . . . . (305) 486 2983
ABBS   Byte Shop, Miami, FL . . . . (305) 261 3639
ABBS   Calvary Mission Church, Minneapolis, MN. . . . . .! (612) 471 0252
ABBS   CCNJ, Pompton Plains, NJ . . . . (201) 835 7228
ABBS   Century Next Computers, St. Louis, MO . . . . (314) 442 6502
ABBS   Charlotte, NC . . . .! (704) 364 5254
ABBS   Cleveland, OH . . . . (216) 779 1338
ABBS   CODE, Glen Ellyn IL . . . . (312) 537 7063 *24
ABBS   Colortron Computer, WI . . . . (414) 637 9990 *24
ABBS   Compumart, Ottawa, Ontario, Canada . . . . (613) 725 2243
ABBS   Computerland, Fremont, CA . . . . (415) 794 9314
ABBS   Computer Corner, Amarillo, TX . . . . (806) 355 5610
ABBS   Computer Conspiracy, Santa Monica, CA . . . . (213) 829 1140
ABBS   Computer Crossroads, Columbia, MD . . . . (301) 730 0922
ABBS   Computer Lab, Memphis, TN . . . . (901) 761 4743
ABBS   Computer Room, Kalamazoo, MI . . . . (616) 382 0101
ABBS   Computer Store, Toledo, OH . . . . (419) 531 3845
ABBS   Dallas Info Board . . . . (214) 248 4539
ABBS   Denver, CO . . . . (303) 759 2625
ABBS   Detroit, MI . . . . (313) 477 4471
ABBS   Downer's Grove, IL . . . . (312) 964 7768
ABBS   Electro-Mart, Spokane, WA . . . . (509) 534 2419 *24
ABBS   Fort Walton Beach, Destin, FL . . . . (904) 243 1257
ABBS   Gamemaster, Chicago, IL . . . . (312) 475 4884 *24
ABBS   Hayward, CA . . . . (415) 881 5662
ABBS   Illinois Microcomputer, Naperville, IL . . . . (312) 420 7995
ABBS   Ketchikan, AK . . . . (907) 225 6789
ABBS   Louisville, KY . . . . (502) 245 7811 *24
ABBS   Madam Bokeatha Society, Houston, TX . . . . (713) 455 9502
ABBS   Memphis, TN . . . . (901) 725 5691
ABBS   Michigan Apple-Fone, Southfield, MI . . . . (313) 357 1422
ABBS   Newport Beach, CA . . . .! (714) 645 5256 *24
ABBS   Oak Brook Computer, Oak Brook, IL . . . . (312) 941 9009
ABBS   Omaha, NE . . . . (402) 339 7809
ABBS   PCnet, San Francisco, CA . . . . (415) 863 4703 *24
ABBS   Pacific Palisades, Los Angeles, CA . . . . (213) 459 6400
ABBS   Peoria, IL . . . . (309) 692 6502
ABBS   Phoenix, AZ . . . . (602) 898 0891
ABBS   Pirates Cove, Long Island, NY . . . . (516) 698 4008
ABBS   Rogers Park, Chicago, IL . . . . (312) 973 2227
ABBS   San Antonio, TX . . . . (512) 737 0214 *24
ABBS   Software Sorcery, Herndon, VA . . . .! (703) 471 0610
ABBS   South of Market, San Francisco, CA . . . . (415) 469 8111 -so
ABBS   Spokane, WA . . . . (509) 456 8900
ABBS   St. Louis, MO . . . . (314) 838 7784 *24
ABBS   Teledunjon I, Dallas, TX . . . . (817) 469 1626
ABBS   Teledunjon II, Dallas, TX . . . . (214) 530 0858
ABBS   Teledunjon III, Dallas, TX . . . . (214) 960 7654

ABBS	The Moon, Dallas, TX	(214) 931 3437 *24
ABBS	Turnersville, NJ	(609) 228 1149
ABBS	Vancouver, B.C.	(604) 437 7001
ABBS	Vermont, Essex Junction, VT	(802) 879 4981 *24
ABBS	VIBBS, Nashua, NH	(603) 888 6648
ABBS	West Palm Beach, FL	(305) 848 3802
ABBS	Rob Roy Computer, Yakima, WA	(509) 575 7704
ABBS	Youngs Elect Svc., College Station, TX	(713) 693 3462 *24
ABBS	#X, Atlanta, GA	(404) 256 1549
A-C-C-E-S-S	Annapolis, MD	(301) 267 7666 *24
A-C-C-E-S-S	Olympia, WA	(206) 866 9043 *24
A-C-C-E-S-S	Phoenix, AZ ....#1	(602) 996 9709 *24
A-C-C-E-S-S	Phoenix, AZ ....&	(602) 957 4428 *24
A-C-C-E-S-S	Phoenix, AZ	(602) 274 5964
A-C-C-E-S-S	Scotsdale, AZ	(602) 998 9411 *24
A-C-C-E-S-S	Wyckoff, NJ	(201) 891 7441 *24
AMIS	A.R.C.A.D.E. Sterling Heights, MI	(313) 978 8087 *24
AMIS	Chicago, IL	(312) 789 3610 *24
AMIS	APOGEE Miami, FL	(305) 238-1231 -rb
AMIS	GRAFEX, Cupertino, CA	(408) 253 5216
AMIS	G.R.A.S.S. Grand Rapids, MI	(616) 241 1971 *24
AMIS	IBBBS San Jose, CA	(408) 298 6930
AMIS	M.A.C.E. Detroit, MI ....#1	(313) 868 2064 *24
AMIS	Magic Lantern, Madison, WI	(608) 251 8538
AMIS	Starbase 12 Philadelphia, PA	(617) 876 4885
AMIS	T.A.B.B.S. Sunnyvale, CA	(408) 942 6975
ARMUDIC	Washington, D.C. ....#1	(202) 276 8342
ARMUDIC	Computer Age, Baltimore, MD	(301) 587 2132
BBS	Annandale, VA	(703) 978 9754
BBS	B.R., Los Angeles, CA	(213) 394 5950 *24
BBS	Computer Applications Co., Poland, OH	(216) 757 3711
BBS	Electronic Exchange, Chicago, IL	(312) 541 6470 *24
BBS	Homestead, FL	(305) 246 1111
BBS	IBM PC Billings, MT	(406) 656 9624
BBS	IBM PCUG Annandale, VA	(703) 560 0979
BBS	Living Videotext, Menlo Park, CA	(415) 327 8876 *24
BBS	Pensacola, FL	(904) 477 8783
BBS	SUE Milwaukee, WI	(414) 483 4578
BBS-80	Cincinnati, OH	(513) 244 2983
BBS-80	DALTRUG, Dallas TX	(214) 235 8784 *24
BULLET-80	Akron, OH	(216) 645 0827 *24
BULLET-80	Boston, MA ....&	(617) 266 7789 *24
BULLET-80	Chesterland, OH	(216) 729 2769
BULLET-80	Clarks Summit, PA	(717) 586 2112
BULLET-80	Danbury, CT ....#1	(203) 744 4644

BULLET-80 Fayetteville, GA .... (404) 461 9689
BULLET-80 Fawkins, TX .... (214)769 3036
BULLET-80 Houston, TX .... (713) 331 2599
BULLET-80 Ironton, OH .... (614) 532 6920
BULLET-80 Lancaster, CA .... (805) 947 9925
BULLET-80 Langhorne, PA .... (215) 364 2180
BULLET-80 Littlefield, TX .... (806) 385 6843
BULLET-80 Mt. Clemens, MI .... (313) 465 9531
BULLET-80 Orange County, Anaheim, CA............. (714) 952 2110
BULLET-80 Poughkeepsie, NY .... (914) 278 2375 -so
BULLET-80 Redwood City, CA .... (415) 367 1339 *24
BULLET-80 Riverside, CA .... (714) 359 3189
BULLET-80 San Jose, CA .... (408) 241 0769
BULLET-80 Seymour, CT .... (203) 888 7952
BULLET-80 Springfield, IL......................... (217) 529 1113
BULLET-80 Tulsa, OK .... (918) 749 0059 *24
BULLET-80 Tyler, TX............................. (214) 595 4217

CBBS AMRAD, Washington, DC.................... (703) 734 1387 *24
CBBS Atlanta, GA .... (404) 394 4220 *24
CBBS Baton Rouge, LA .... (504) 273 3116 *24
CBBS Bloomington, IN .... (812) 334 2522
CBBS Boston, MA .... (617) 646 3610 *24
CBBS Cedar Rapids, IA .... (319) 364 0811
CBBS Chicago, IL...............................#1 (312) 545 8086 *24
CBBS Corpus Christi, TX .... (512) 855 1512
CBBS CPEUB/ICST Gaithersburg, MD .... (301) 948 5717
CBBS Lambda, Berkeley, CA .... (415) 658 2929 -so
CBBS Lawrence General Hospital, Boston, MA.......... (617) 683 2119
CBBS LICA LIMBS, Long Island, NY .... (516) 561 6590 *24
CBBS London, England ...... (European standard) ...... (044) 1 399 2136
CBBS Long Island, NY............................. (516) 334 3134 *24
CBBS MAUDE Milwaukee, WI .... (414) 241 8364 *24
CBBS NW, Portland, OR.......................... (503) 646 5510 *24
CBBS PACC, Pittsburgh, PA .... (412) 822 7176 *24
CBBS Prince George, B.C., Canada .... (604) 562 9515
CBBS Proxima, Berkeley, CA .... (415) 357 1130
CBBS RAMS, Rochester, NY .... (716) 244 9531
CBBS Richfield, MN .... (612) 869 5780
CBBS TSG, Tucson, AZ............................ (602) 746 3956 *24
CBBS Vancouver, BC, Canada .... (604) 687 2640 *24
CBBS Waco, TX .... (817) 776 1375

COMM-80 OCTUG .... (714) 530 8226
COMM-80 Orange County, Fullerton, CA .... (714) 530 8226
COMM-80 Queens, NY .... (212) 897 3392 *24

CONNECTION-80 Centereach, NY .... (516) 588 5836
CONNECTION-80 Denver, CO .... (303) 690 4566 *24

CONNECTION-80 Escondido, CA . . . .! (714) 746 6265
CONNECTION-80 Fremont, CA . . . . (415) 651 4147 *24
CONNECTION-80 Gaithersburg, MD . . . . (301) 840 8588 *24
CONNECTION-80 Great Neck, NY . . . . (516) 482 8491 *24
CONNECTION-80 Lansing, MI . . . . (517) 339 3367
CONNECTION-80 Little Rock, AS . . . . (501) 372 0576
CONNECTION-80 Manhattan, NY . . . . (212) 991 1664
CONNECTION-80 Orlando, FL . . . . (305) 644 8327 *24
CONNECTION-80 PAUG, Portland, OR . . . . (503) 281 7653
CONNECTION-80 Peterborough, NH . . . . (603) 924 7920
CONNECTION-80 Tulsa, OK . . . . (918) 747 1310 *24
CONNECTION-80 W. Mich. Micro Group, MI . . . . (616) 457 1840 *24
CONNECTION-80 Winter Garden, FL . . . . (305) 877 2829 *24
CONNECTION-80 Woodhaven, NY . . . . (212) 441 3755 *24
CONNECTION-80 Tampa, FL . . . . (813) 977 0989

CONFERENCE-TREE #2, San Francisco, CA . . . . (415) 928 0641
CONFERENCE-TREE #3, Hayward, CA . . . . (415) 538 3580
CONFERENCE-TREE #4, Santa Monica, CA . . . . (213) 394 1505
CONFERENCE-TREE Anchorage, AK . . . . (907) 344 5251
CONFERENCE-TREE Flagship, Denville, NJ . . . . (201) 627 5151 *24
CONFERENCE-TREE Kelp Bed, Los Angeles, CA . . . . . . (213) 372 4800
CONFERENCE-TREE Minneapolis, MN . . . . (612) 227 0307
CONFERENCE-TREE ?, New Jersey . . . . (201) 627 5151
CONFERENCE-TREE Victoria, TX . . . . (512) 578 5833

DIAL-YOUR-MATCH #1   . . . . (213) 842 3322 -so
DIAL-YOUR-MATCH #3   . . . . (912) 233 0863 -so
DIAL-YOUR-MATCH #4   . . . . (213) 704 9819 -so
DIAL-YOUR-MATCH #7   . . . . (212) 456 2528 -so
DIAL-YOUR-MATCH #9   . . . . (213) 345 1047 -so
DIAL-YOUR-MATCH #11 . . . .! (714) 242 1882

FORUM-80 Albany, NY . . . . (518) 785 8478
FORUM-80 Augusta, GA . . . . (803) 279 5392
FORUM-80 Charleston, SC . . . . (803) 552 1612 *24
FORUM-80 Cleveland, OH . . . .& (216) 486 4176
FORUM-80 #1, Denver, CO . . . . (303) 341 0636 *24
FORUM-80 #2, Denver, CO . . . . (303) 399 8858 *24
FORUM-80 El Paso, TX . . . . (915) 755 1000 *24
FORUM-80 Ft. Lauderdale, FL . . . . (305) 772 4444 *24
FORUM-80 Hull, England . . . .(011) 44 482 859169
FORUM-80 Kansas City, MO . . . .#1 & (816) 861 7040 *24
FORUM-80 Kansas City, MO . . . .& (816) 931 9316
FORUM-80 Las Vegas, NV . . . . (702) 362 3609 *24
FORUM-80 Linden, NJ . . . . (201) 486 2956 *24
FORUM-80 Medford, OR . . . . (503) 535 6883 *24
FORUM-80 Medical, Memphis, TN . . . . (901) 276 8196 *24
FORUM-80 Monmouth, Brielle, NJ . . . . (201) 528 6623 *24

```
FORUM-80  Montgomery, AL . . . . . . . . . . . . . . . . . . . . . . .    (205) 272 5069
FORUM-80  Nashua, NH . . . . . . . . . . . . . . . . . . . . . . . . . . .    (603) 882 5041
FORUM-80  Orange County, Anaheim, CA . . . . . . . . . . . . .    (714) 545 9549
FORUM-80  Pontiac, MI . . . . . . . . . . . . . . . . . . . . . . . . . . .    (313) 335 8456
FORUM-80  Prince William County, VA . . . . . . . . . . . . . .    (703) 670 5881  *24
FORUM-80  San Antonio, TX . . . . . . . . . . . . . . . . . . . . . . .    (512) 340 6720
FORUM-80  Seattle, WA . . . . . . . . . . . . . . . . . . . . . . . . . . .    (206) 723 3282
FORUM-80  Sierra Vista, AZ . . . . . . . . . . . . . . . . . . . . . . . .    (602) 458 3850  *24
FORUM-80  Shreveport, LA . . . . . . . . . . . . . . . . . . . . . . . .    (318) 631 7107  *24
FORUM-80  Westford, MA . . . . . . . . . . . . . . . . . . . . . . . . .    (617) 692 3973
FORUM-80  Wichita, KA . . . . . . . . . . . . . . . . . . . . . . . . .&    (316) 682 2113  *24
FORUM-80  Wichita Falls, TX . . . . . . . . . . . . . . . . . . . . . .    (817) 855 3916

HBBS  Denver, CO . . . . . . . . . . . . . . . . . . . . . . . . . . . . .    (303) 343 8401  *24
HBBS  El Paso, TX . . . . . . . . . . . . . . . . . . . . . . . . . . . . .    (915) 592 1910
HBBS  Oklahoma City, OK . . . . . . . . . . . . . . . . . . . . . . . .    (405) 848 9329  *24
HBBS  San Fernando Valley . . . . . . . . . . . . . . . . . . . . . . .    (213) 366 4837

MCMS  C.A.M.S. Chicago, IL . . . . . . . . . . . . . . . . . . . . .#1    (312) 927 1020  *24
MCMS  J.A.M.S. Lockport, IL . . . . . . . . . . . . . . . . . . . . .    (815) 838 1020  *24
MCMS  L.A.M.S. Round Lake, IL . . . . . . . . . . . . . . . . . . .    (312) 740 9128
MCMS  Message-82, Chicago, IL . . . . . . . . . . . . . . . . . . . .    (312) 622 4442  *24
MCMS  Metro West Database, Chicago, IL . . . . . . . . . . . . . .    (312) 260 0640  *24
MCMS  NC Sofware, Minneapolis, MN . . . . . . . . . . . . . . .    (612) 533 1957  *24
MCMS  WACO Hot Line, Schaumburg, IL . . .    pvt    . . .    (312) 351 4374  *24

MICRO-COM  Computer City, Weston, CT . . . . . . . . . . . .    (203) 227 1829
MICRO-COM  ? . . . . . . . . . . . . . . . . . . . . . . . . . . . . . . . .    (703) 780 0610

MOUSE-NET  Nashua, NH . . . . . . . . . . . . . . . . . . . . . . . .    (603) 673 9476  *24
MOUSE-NET  Orlando, FL . . . . . . . . . . . . . . . . . . . . . . . .    (305) 277 0473  *24

MSG-80  Everett, WA . . . . . . . . . . . . . . . . . . . . . . . . . . .    (206) 334 7394

NESSY  Chicago, IL . . . . . . . . . . . . . . . . . . . . . . . . . . . .#1    (312) 773 3308
NESSY  Terry and Gwen's, Palatine, IL . . . . . . . . . . . . . . .    (312) 289 6393

NET-WORKS  Apple Grove, Dallas, TX . . . . . . . . . . . . . . .    (214) 644 5197
NET-WORKS  Apple Shack, Dallas, TX . . . . . . . . . . . . . . .    (214) 644 4781  *24
NET-WORKS  Armadillo, Grand Fork, ND . . . . . . . . . . . .    (701) 746 4959
NET-WORKS  Aurora, CO . . . . . . . . . . . . . . . . . . . . . . . .    (303) 343 8401  *24
NET-WORKS  Big Apple, Miami, FL . . . . . . . . . . . . . . . . .    (305) 948 8000
NET-WORKS  C.A.M.S., Decatur, IL . . . . . . . . . . . . . . . .    (217) 429 5541
NET-WORKS  CLAH, Arlington Hts., IL . . . . . . . . . . . . . .    (312) 255 6489
NET-WORKS  Coin Games, Los Angeles, CA . . . . . . . . . . .    (213) 336 5535
NET-WORKS  COMM Center NW3NAGAD, Laurel, MD    (301) 953 1110
NET-WORKS  Computer City, Providence, RI . . . . . . . . . .    (401) 331 8450  *24
NET-WORKS  Computer Emporium, Des Moines, IA . . . .    (515) 279 8863
NET-WORKS  Computer Emporium, San Jose, CA . . . . . .    (408) 227 0227
NET-WORKS  Computer Pro, Ft. Worth, TX . . . . . . . . . . .    (817) 732 1787
```

NET-WORKS Computer Station, St. Louis, MO . . . . (314) 432 7120
NET-WORKS Computer World, Los Angeles, CA . . . . (213) 859 0894 *24
NET-WORKS Crescent City, Baton Rouge, LA . . . . (504)454 6688
NET-WORKS Dallas, TX . . . . (214) 361 1386 *24
NET-WORKS Dayton, OH . . . . (513) 223 3672
NET-WORKS Eclectic Computer Sys., Dallas, TX . . . . (214) 239 5842
NET-WORKS Encino, CA . . . . (213) 345 3670
NET-WORKS Granite City, IL . . . . (618) 877 2904
NET-WORKS Greenfield, IN . . . . (317) 326 3833 *24
NET-WORKS Hacker-net, Dallas, TX . . . . (214) 824 7160
NET-WORKS Hawaii . . . . (808) 521 7312
NET-WORKS Info-Net, Costa Mesa, CA . . . . (714) 545 7359
NET-WORKS MAGIE, Galesburg, IL . . . . (309) 342 7178
NET-WORKS Magnetic Fantasies, Los Angeles, CA . . . . . (213) 388 5198
NET-WORKS Montreal, Canada . . . . (514) 937 2188 *24
NET-WORKS Pirates' Harbor, Boston, MA . . . . (617) 738 5051
NET-WORKS Pirate's Inn . . . . (213) 454 3075
NET-WORKS Pirate's Nest, Weston, CT . . . . (203) 227 1829
NET-WORKS Pirate's Ship, Chicago, IL . . . . (312) 935 2933 *24
NET-WORKS Portsmouth, NH . . . . (603) 436 3461
NET-WORKS Softworx, West Los Angeles, CA . . . . (213) 473 2754
NET-WORKS Sparklin' City, Corpus Christi, TX . . . . (512) 882 6569
NET-WORKS St. Louis, MO . . . . (314) 781 1308
NET-WORKS Warlock's Castle St. Louis, MO . . . . (618) 345 6638
NET-WORKS Winesap, Dallas, TX . . . . (214) 824 7455

ONLINE CDC, San Diego, CA . . . . ! (714) 452 6012
ONLINE Houston, TX . . . . (713) 528 7025 *24
ONLINE Dickinsons Movie Guide, Missions, KS . . . . (913) 432 5544
ONLINE Indianapolis, IN. ID# = GUES,
pswd = pass . . . . (317) 787 9881 *24
ONLINE ON-LINE Computer Center, KC, MO . . . . (913) 341 7987
ONLINE Saba, San Diego, CA . . . . ! (714) 291 5229
ONLINE Santee, CA .   ID# = GUEST, pswd = PASS   . (714) 561 7271 *24

PASBBS Bellflower, CA . . . . (213) 531 1057
PASBBS Torrance, CA . . . . #1 (213) 516 7089 *24

PBBS Co-operative Comp Svc, Palatine, IL . . . . (312) 359 9450 *24

PET BBS S.E.W.P.U.G., Racine, WI . . . . (414) 554 9520 *24
PET BBS Commodore Comm., Lake St. Louis, MO . . . . .! (314) 625 4576 *24

PMS—**IF**, Anaheim, CA . . . . (714) 772 8868 *24
PMS—Anchorage, AK . . . . (907) 344 8558
PMS—Apple Bits, Kansas City, MO . . . . (913) 341 3502 *24
PMS—Apple Guild, Weymouth, MA . . . . (617) 767 1303 *24
PMS—Arlington Heights, IL . . . . (312) 870 7176 *24
PMS—Baltimore, MD . . . . (301) 764 1995 *24

PMS—Campbell, CA .... (408) 370 0873 *24
PMS—Century 23, Las Vegas,NV .... ! (702) 878 9106 *24
PMS—Chicago, IL .... (312) 373 8057 *24
PMS—Cincinnati, OH.............................. (513) 671 2753
PMS—Computer City, Danvers, MA .... (617) 774 7516
PMS—Ellicott City, MD .... (301) 465 3176
PMS—Escondido, CA.............................. (714) 746 0667
PMS—Ft. Smith Computer Club, Ft. Smith, AK .... ! (501) 646 0197
PMS—Gulfcoast, Freeport, TX .... (713) 233 7943 *24
PMS—Indianapolis, IN............................. (317) 787 5486 *24
PMS—Lakeside, CA. (type PMS to activate) .... (714) 561 7271 *24
PMS—Los Angeles, CA .... (213) 334 7614 *24
PMS—Massillon, OH .... (216) 832 8392 *24
PMS—McGraw-Hill Books, New York, NY .... (212) 997 2488
PMS—Mesa Systems, San Diego, CA .... ! (714) 271 8613
PMS—Minneapolis, MN............................ (612) 929 6699 *24
PMS—Mission Valley, CA .... (714) 295 8280
PMS—I.A.C., Lake Forest, IL .... (312) 295 6926 *24
PMS—O.A.C., Woodland Hills, CA .... (213) 346 1849 *24
PMS—Pikesville, MD .... (301) 653 3413
PMS—Pleasanton, CA .... (415) 462 7419 *24
PMS—Portola Valley, CA .... (415) 851 3453 *24
PMS—RAUG, Akron, OH........................... (216) 867 7463 *24
PMS—Richland, WA .... (509) 943 6502 *24
PMS—Rutgers Univ. Microlab, Piscataway, NJ .... (201) 932 3887
PMS—San Diego, CA .... (714) 582 9557 *24
PMS—Santa Clara, CA............................. (408) 554 9036
PMS—Santa Cruz, Aptos, CA .... (408) 688 9629 *24
PMS—Santee, CA ....#1 (714) 561 7277 *24
PMS—Shrewsbury, NJ .... (201) 747 6768
PMS—Software Ultd, Kenmore, WA.................. (206) 486 2368 *24
PMS—Tampa, FL .... (down temp)
PMS—Your Computer Connection, KS Cty, MO .... (913) 381 1021

PSBBS Baltimore, MD .... (301) 994 0399 *24
PSBBS Washington, DC .... (202) 337 4694 *24

RATS Systems................................#1 (201) 887 8874
RATS Pequannock, NJ .... (201) 696 8647
RATS Little Falls, NJ .... (201) 785 3565
RATS Homewood, IL .... (312) 957 3924
RATS Wenonah, NJ.............................. (609) 468 5293
RATS Wenonah, NJ #2 .... (609) 468 3844

RBBS Big Top, Milwaukee, WI .... (414) 259 9475

RCP/M AABB New York, NY .... (212) 787 5520
RCP/M AIMS Hinsdale, IL........................! (312) 789 0499
RCP/M Arlington, VA............................! (703) 536 3769

```
RCP/M  CBBS ANAHUG, Anaheim, CA .............  (714) 774 7860
RCP/M  CBBS CP/M Net Simi Valley, CA ............  (805) 527 9321
RCP/M  CBBS Columbus, OH ......................  (614) 272 2227 *24
RCP/M  CBBS Frog Hollow, Vancouver, BC, CN .......!  (604) 873 4007 *24
RCP/M  CBBS HUG, Chicago, IL ...................  (312) 671 4992 *24
RCP/M  CBBS Pasadena, CA ......................  (213) 799 1632 *24
RCP/M  CBBS RLP, MacLean, VA ..................!  (703) 524 2549 *24
RCP/M  CBBS Sacramento, CA ....................!  (916) 483 8718 *24
RCP/M  CBBS Vancouver, BC, Canada ..............  (604) 687 2640 *24
RCP/M  Chuck Forsberg, OR .....................!  (503) 621 3193
RCP/M  Collossal Oxgate, ??.....................!  (408) 263 2588
RCP/M  CUG-NOTE, Denver, CO ...................!  (303) 781 4937 *24
RCP/M  CUG-NODE, PA State College...............!  (814) 238 4857 *24
RCP/M  Detroit, MI .............................  (313) 584 1044  -rb
RCP/M  Geneseo, IL.............................!  (309) 944 5455
RCP/M  HAPN Hamilton, Ontario, CN................!  (416) 335 6620 *24
RCP/M  IBM-PC, Niles, IL .......................&!  (312) 647 7636 *24
RCP/M  Logan Square, Chicago, IL ..................  (312) 252 2136
RCP/M  MCBBS Keith Petersen, Royal Oak, MI .........  (313) 759 6569  -rb
RCP/M  MCBBS Ken Stritzel, Flanders, NJ ............!  (201) 584 9227 *24
RCP/M  MCBBS Superbrain, Lexington, MA ........$&  (617) 862 0781 *24
RCP/M  MCBBS TCBBS Dearborn, MI...............  (313) 846 6127 *24
RCP/M  Mississauga, Toronto, Ontario, CN..........$&  (416) 826 5394 *24
RCP/M  Mississauga HUG, Toronto, Ontario, CN ....$&  (416) 826 5394 *24
RCP/M  NEI, Chicago, IL .........................  (312) 949 6189
RCP/M  Palatine, IL ............................&  (312) 359 2553 *24
RCP/M  Piconet Oxgate, ?? ......................  (415) 965 4097 *24
RCP/M  RBBS Allentown, PA ....................  (215) 398 3973 *24
RCP/M  RBBS ANAHUG, Anaheim, CA ..............  (714) 774 7860 *24
RCP/M  RBBS Arvada Elect., Colorado Springs, CO. ....  (303) 634 1158 *24
RCP/M  Boulder, CO ...........................  (303) 499 9169
RCP/M RBBS  Bethesda, MD .......................!  (301) 229 3196
RCP/M RBBS  Comp. Tech. Assoc., El Paso, TX .......!  (915) 533 2202 *24
RCP/M RBBS  Computerized Services, Tampa, FL .......  (813) 839 6746
RCP/M RBBS  Computron, Edmonton, Alberta, Can ....  (403) 482 6854 *24
RCP/M RBBS  Cranford, NJ ........................  (201) 272 1874
RCP/M RBBS  DataTech, San Carlos, CA .............  (415) 595 0541
RCP/M RBBS  Edmonton, Alberta, Canada ............  (403) 454 6093 *24
RCP/M RBBS  El Paso, TX .........................  (915) 598 1668
RCP/M RBBS  Fort Mill, SC ........................  (803) 548 0900 *24
RCP/M RBBS  GFRN Dta Exch., Garden Grove, CA& .$&  (714) 534 1547 *24
RCP/M RBBS  GFRN Dta Exch., Palos Verdes, CA....$&  (213) 541 2503 *24
RCP/M RBBS  Grafton, VA.........................!  (804) 898 7493
RCP/M RBBS  Huntsville, AL .......................  (205) 895 6749  -rb
RCP/M RBBS  Hyde Park, IL .......................  (312) 955 4493
RCP/M RBBS  Laurel, MD .........................!  (301) 953 3753 *24
RCP/M RBBS  Larkspur, CA .......................  (415) 461 7726 *24
```

RCP/M RBBS Long Island, NY .... (516) 698 8619 -rb
RCP/M RBBS Marin County, CA ....! (415) 383 0473 *24
RCP/M RBBS MUG, Mission, KS...................& (913) 362 9583 *24
RCP/M RBBS Napa Valley, CA .... (707) 253 1523
RCP/M RBBS New York, NY....................... (516) 791 1767
RCP/M RBBS Ocean, NJ..........................& (201) 775 8705
RCP/M RBBS San Jose Oxgate, San Jose, CA..........! (408) 287 5901 *24
RCP/M RBBS Surrey, Vancouver, BC, CN ....! (604) 584 2643 *24
RCP/M RBBS Paul Bogdanovich, NJ ....! (201) 747 7301
RCP/M RBBS Rochester, NY .... (716) 223 1100 *24
RCP/M RBBS San Diego, CA ....$&! (714) 273 4354 *24
RCP/M RBBS Sofwaire Store, Los Angeles, CA .... (213) 479 3189 *24
RCP/M RBBS Software Tools, Australia .... (02) 997 1836
RCP/M RBBS South Florida .... (305) 255 6027
RCP/M RBBS Southfield, MI...................... (313) 559 5326 *24
RCP/M RBBS Thousand Oaks, CA .... (805) 496 9522 *24
RCP/M RBBS Westland, MI .... (313) 729 1905 -rb
RCP/M RBBS Woodstock, NY.....................! (914) 679 8734
RCP/M RBBS Yelm, WA .... (206) 458 3086 -rb
RCP/M Silicon Valley, CA ....! (408) 246 5014 *24
RCP/M SJBBS Bearsville, NY .... (914) 679 6559 -rb
RCP/M SJBBS Johnson City, NY .... (607) 797 6416
RCP/M TRS-80 Chicago, IL........................ (312) 949 6189
RCP/M Terry O'Brien, Vancouver, BC, Canada......... (604) 584 2543

Remote Northstar Atlanta, GA ....#1 (404) 926 4318 *24
Remote Northstar Denver, CO....................... (303) 444 7231
Remote Northstar Largo, FL .... (813) 535 9341 *24
Remote Northstar NASA, Greenbelt, MD .... (301) 344 9156
Remote Northstar Santa Barbara, CA................. (805) 682 7876
Remote Northstar Santa Barbara, CA................. (805) 964 4115
Remote Northstar Virginia Beach, VA .... (804) 340 5246

ST80-CC Lance Micklus, Inc., Burlington, VT.........#1 (802) 862 7023 *24
ST80-PBB Monroe Camera Shop, Monroe, NY .... (914) 782 7605

TCBBS B.A.M.S. New York, N.Y.................... (212) 362 1040 *24
TCBBS Leigh's Computer World,NY .... (212) 879 7698
TCBBS W.E.B.B. New York, NY ....#1 (212) 799 4649

TRADE-80 Ft. Lauderdale, FL ....#1 (305) 525 1192
TRADE-80 Omaha, NE .... (402) 292 6184
TRADE-80 Erie, PA.............................. (814) 898 2952 *24

Miscellaneous or Unknown System Types

Adventure BBS	(516) 621 9296	
Alpha, Tampa, FL . acct# = ABCDOO, pwd = TRYIT	(813) 251 4095	*24
Apollo's Chariot, Apollo, FL	(813) 645 3669	
ARBB Seattle, WA	(206) 546 6239	
Aunt Dru's X Rated, North Whales, PA	(215) 855 3809	
Aviators Bulletin Board, Sacramento, CA	(916) 393 4459	
Bathroom Wall BBS, San Antonio, TX !	(512) 655 8143	
Baton Rouge Data System, Baton Rouge, LA	(504) 926 0181	
Boston Information Exchange, Boston, MA &	(617) 423 6985	*24
Bronx BBS, NY	(212) 933 9459	
BR's BBS Santa Monica, CA	(213) 394 5950	
Bradley Computer BBS	(813) 734 7103	
BSBB Tampa, FL	(813) 885 6187	
Capital City BBS, Albany, NY	(518) 346 3596	*24
Carrier 2 Alexandria, VA	(703) 823 5210	
C-HUG Bulletin Board, Fairfax, VA	(703) 360 3812	*24
Compusystems, Columbia, SC	(803) 771 0922	
Computer Arts Message System, Newhall, CA	(805) 255 6445	
Computer Conspiracy	(213) 829 1140	
Corsair, Goldcoast, FL	(305) 968 8653	
CoxCo, Arvada, CO	(303) 423 5001	*24
Databoard BBS, Atherton, CA !	(415) 367 7638	*24
Datamate, Canoga Park, CA !	(213) 998 7992	-so
Dimension-80 Orange, CA	(714) 974 9788	
Dragon's Game System (pass = DRAGON)	(213) 428 5206	
Electric Line Connection, Sherman Oaks, CA	(213) 789 9512	
Experimental-80 Kansas City, MO	(913) 676 3613	
FBBS Skokie, IL #1	(312) 677 8514	
Greene Machine, WPB, FL	(305) 965 4388	-so
Hermes-80 Allentown, PA	(215) 434 3998	
HEX Silver Spring, MD %	(301) 593 7033	*24
HMS Horny Message System, Oakland, CA	(415) 845 2079	-so
IAC Message Base, Menlo Park, CA	(415) 367 1339	*24
INFO-NET Foster City, CA	(415) 349 3126	
INFOEX-80 West Palm Beach, FL	(305) 683 6044	*24
JCTS Redmond, WA	(206) 883 0403	*24
Kinky Kumputer, San Francisco, CA	(415) 647 9524	-so
Kluge Computer $&	(213) 947 8128	*24
L.A. Interchange, Los Angeles, CA	(213) 631 3186	*24
Lehigh Press BB, PA #1	(215) 435 3388	
Long Beach Community Computer	(213) 591 7239	*24
Mail Board-82, Seattle, WA	(206) 527 0897	*24
Market 80, Kansas City, MO	(816) 931 9316	
MARS/RP Rogers Park, IL	(312) 743 8176	*24
Micro-80 West Palm Beach, FL	(305) 686 3695	
Micro Informer	(813) 884 1506	
Midwest, St. Louis, MO	(314) 227 4312	-so
Mini-Bin Seattle, WA	(206) 762 5141	*24

```
NBBS Norfolk, VA.................................!  (804) 444 3392
New England Comp. Soc., Maynard, MA ..............  (617) 897 0346
New Jersey TELECOM ...........................#1  (201) 635 0705  *24
NMS Natick, MA ..................................  (617) 653 4282  *24
North Largo, FL..................................  (813) 535 9341  *24
North Orange County Computer Club ................  (714) 633 5240
Novation Co., Los Angeles, CA  ....   pass = CAT   ....  (213) 881 6880
Nybbles-80 Elmsford, NY ..........................  (914) 592 5385
OARCS Portland, Oregon...........................  (503) 641 2798
Oracle North Hollywood, CA .......................  (213) 980 6743  -so
ORACLE Classified System, Austin, TX ..............  (512) 346 4495
Orange County Data Exchange, Garden Grove, CA ......  (714) 537 7913
Personal Message System-80, Deerfield Bch, FL ..........  (305) 427 6300  *24
PHOTO-80, Haledon, NJ ...........................  (201) 790 6795
Potomac Micro Magic Inc., Falls Church, VA ............  (703) 379 0303  *24
Remote Apple Jackson, MS.........................  (601) 992 1918  *24
SATUG BBS, San Antonio, TX......................!  (512) 494 0285
Scanboard 80, Atlanta, GA .........................  (404) 457 8384
Seacomm-80 Seattle, WA ..........................  (206) 763 8879  *24
SIGNON Reno, NV ........   pswd = FREE   ........  (702) 826 7234
SISTER Staten Island, NY..........................  (212) 442 3874  *24
SLAMS St. Louis, MO..............................  (314) 839 4307
SMBBS New York, NY .............................  (212) 884 5408
Sunrise Omega-80, Oakland, CA ....................  (415) 452 0350
Talk-80 ROBB, Portsmouth, VA ....................  (804) 484 9636
TCUG BBS, Washington, DC........................  (703) 451 8475  *24
Telcom 7 New Fairfield, CT.........................  (203) 746 5763  *24
Telemessage-80, Atlanta, GA .......................  (404) 962 0616
Westside Download, Detroit, MI ....................  (313) 533 0254
Wild goose board, ? ...............................  (813) 988 7400
XBBS Hamilton, OH ..............................  (513) 863 7681  *24
ZBBS Minneapolis, MN............................  (612) 426 9028
```

Public Access Message Systems

LISTING 2: By Area Code

201

```
A-C-C-E-S-S Wyckoff, NJ ..........................  (201) 891 7441  *24
ABBS Apple Group N.J., Piscataway, NJ ..............  (201) 968 1074
ABBS Apple-Mate, New York, NY....................  (201) 864 5345
ABBS CCNJ, Pompton Plains, NJ ....................  (201) 835 7228
CONFERENCE-TREE Flagship, Denville, NJ ..........  (201) 627 5151  *24
```

CONFERENCE-TREE ?, New Jersey . . . . (201) 627 5151
FORUM-80 Linden, NJ . . . . (201) 486 2956 *24
FORUM-80 Monmouth, Brielle, NJ . . . . (201) 528 6623 *24
New Jersey TELECOM . . . . #1 (201) 635 0705 *24
PHOTO-80, Haledon, NJ . . . . (201) 790 6795
PMS—Rutgers Univ. Microlab, Piscataway, NJ . . . . (201) 932 3887
PMS—Shrewsbury, NJ . . . . (201) 747 6768
RATS Little Falls, NJ . . . . (201) 785 3565
RATS Pequannock, NJ . . . . (201) 696 8647
RATS Systems . . . . #1 (201) 887 8874
RCP/M MCBBS Ken Stritzel, Flanders, NJ . . . . ! (201) 584 9227 *24
RCP/M RBBS Cranford, NJ . . . . (201) 272 1874
RCP/M RBBS Ocean, NJ . . . . & (201) 775 8705
RCP/M RBBS Rutgers, New Brunswick, NJ . . . . (201) 932 3879 *24
RCP/M RBBS Paul Bogdanovich, NJ . . . . ! (201) 747 7301

202

ARMUDIC Washington, DC . . . . #1 (202) 276 8342
PSBBS Washington, DC . . . . (202) 337 4694 *24

203

BULLET-80 Danbury, CT . . . . #1 (203) 744 4644
BULLET-80 Seymour, CT . . . . (203) 888 7952
MICRO-COM Computer City, Weston, CT . . . . (203) 227 1829
NET-WORKS Pirate's Nest, Weston, CT . . . . (203) 227 1829
Telcom 7 New Fairfield, CT . . . . (203) 746 5763 *24

205

FORUM-80 Montgomery, AL . . . . (205) 272 5069
RCP/M RBBS Huntsville, AL . . . . (205) 895 6749 -rb

206

A-C-C-E-S-S Olympia, WA . . . . (206) 866 9043 *24
ABBS Apple Crate I, Seattle, WA . . . . (206) 935 9119
ABBS Apple Crate II, Seattle, WA . . . . (206) 244 5438
ARBB Seattle, WA . . . . (206) 546 6239
FORUM-80 Seattle, WA . . . . (206) 723 3282
JCTS Redmond, WA . . . . (206) 883 0403 *24
MSG-80 Everett, WA . . . . (206) 334 7394
Mail Board-82 Seattle, WA . . . . (206) 527 0897 *24
Mini-Bin Seattle, WA . . . . (206) 762 5141 *24
PMS—Software Unltd, Kenmore, WA . . . . (206) 486 2368 *24

RCP/M RBBS Yelm, WA .... (206) 458 3086 -rb
Seacomm-80 Seattle, WA .... (206) 763 8879 *24

212

Bronx BBS, NY .... (212) 933 9459
COMM-80 Queens, NY .... (212) 897 3392 *24
CONNECTION-80 Manhattan, NY.................... (212) 991 1664
CONNECTION-80 Woodhaven, NY .... (212) 441 3755 *24
DIAL-YOUR-MATCH #7 .... (212) 456 2528 -so
PMS—McGraw-Hill Books, New York, NY .... (212) 997 2488
RCP/M AABB New York, NY .... (212) 787 5520
SISTER Staten Island, NY............................ (212) 442 3874 *24
SMBBS New York, NY .... (212) 884 5408
TCBBS B.A.M.S. New York, NY .... (212) 362 1040 *24
TCBBS Leigh's Computer World, NY .... (212) 879 7698
TCBBS W.E.B.B. New York, NY...................#1 (212) 799 4649

213

ABBS Computer Conspiracy, Santa Monica, CA .... (213) 829 1140
ABBS Pacific Palasades, Los Angeles, CA .... (213) 459 6400
BBS B.R., Los Angeles, CA........................... (213) 394 5950 *24
CONFERENCE-TREE #4, Santa Monica, CA .... (213) 394 1505
CONFERENCE-TREE Kelp Bed, Los Angeles, CA .... (213) 372 4800
Computer Conspiracy .... (213) 829 1140
DIAL-YOUR-MATCH #1 .... (213) 842 3322 -so
DIAL-YOUR-MATCH #4 .... (213) 704 9819 -so
DIAL-YOUR-MATCH #9 .... (213) 345 1047 -so
Datamate, Canoga Park, CA....................#1 (213) 998 7992 -so
Dragon's Game System .... (pass = DRAGON) .... (213) 428 5206
Electric Line Connection, Sherman Oaks, CA .... (213) 789 9512
HBBS San Fernando Valley .... (213) 366 4837
Kluge Computer ....$& (213) 947 8128 *24
L.A. Interchange, Los Angeles, CA .... (213) 631 3186 *24
Long Beach Community Computer .... (213) 591 7239 *24
NET-WORKS Coin Games, Los Angeles, CA .... (213) 336 5535
NET-WORKS Computer World, Los Angeles, CA .... (213) 859 0894 *24
NET-WORKS Encino, CA .... (213) 345 3670
NET-WORKS Magnetic Fantasies, Los Angeles, CA ...... (213) 388 5198
NET-WORKS Pirate's Inn .... (213) 454 3075
NET-WORKS Softworx, West Los Angeles, CA .... (213) 473 2754
Novation CO., Los Angeles, CA ....   pass = CAT   .... (213) 881 6880
Oracle North Hollywood, CA .... (213) 980 6743 -so
PASBBS Bellflower, CA .... (213) 531 1057
PASBBS Torrance, CA ....#1 (213) 516 7089 *24
PMS—Los Angeles, CA .... (213) 334 7614 *24

PMS—O.A.C., Woodland Hills, CA . . . . (213) 346 1849 *24
RCP/M CBBS Pasadena, CA . . . . (213) 799 1632 *24
RCP/M RBBS GRFN Dta Exch., Palos Verdes, CA . . . . $& (213) 541 2503 *24
RCP/RBBS Sofwaire Store, Los Angeles, CA . . . . (213) 479 3189 *24

214

ABBS ABBS, Dallas, TX . . . . (214) 661 2969 *24
ABBS Dallas Info Board . . . . (214) 248 4539
ABBS Teledunjon II, Dallas, TX . . . . (214) 530 0858
ABBS Teledunjon III, Dallas, TX . . . . (214) 960 7654
ABBS The Moon, Dallas, TX . . . . (214) 931 3437 *24
BBS-80 DALTRUG, Dallas, TX . . . . (214) 235 8784 *24
BULLET-80 Hawkins, TX . . . . (214) 769 3036
BULLET-80 Tyler, TX . . . . (214) 595 4217
NET-WORKS Apple Grove, Dallas, TX . . . . (214) 644 5197
NET-WORKS Apple Shack, Dallas, TX . . . . (214) 644 4781 *24
NET-WORKS Dallas, TX . . . . (214) 361 1386 *24
NET-WORKS Eclectic Computer Sys., Dallas, TX . . . . (214) 239 5842
NET-WORKS Hackner-net, Dallas, TX . . . . (214) 824 7160
NET-WORKS Winesap, Dallas, TX . . . . (214) 824 7455

215

Aunt Dru's X Rated, North Whales, PA . . . . (215) 855 3809
BULLET-80 Langhorne, PA . . . . (215) 364 2180
Hermes-80 Allentown, PA . . . . (215) 434 3998
Lehigh Press BB, PA . . . . #1 (215) 435 3388
RCP/M RBBS Allentown, PA . . . . (215) 398 3937 *24

216

ABBS Akron Digital Group, Akron, OH . . . . (216) 745 7855 *24
ABBS Cleveland, OH . . . . (216) 779 1338
BBS Computer Applications Co., Poland, OH . . . . (216) 757 3711
BULLET-80 Akron, OH . . . . (216) 645 0827 *24
BULLET-80 Chesterland, OH . . . . (216) 729 2769
FORUM-80 Cleveland, OH . . . . & (216) 486 4176 *24
PMS—Massillon, OH . . . . (216) 832 8392 *24
PMS—RAUG, Akron, OH . . . . (216) 867 7463 *24

217

BULLET-80 Springfield, IL . . . . (217) 529 1113
NET-WORKS C.A.M.S., Decatur, IL . . . . (217) 429 5541

301

A-C-C-E-S-S Annapolis, MD .... (301) 267 7666 *24
BBS Computer Crossroads, Columbia, MD............. (301) 730 0922
ARMUDIC Computer Age, Baltimore, MD .... (301) 587 2132
CBBS CPEUG/ICST Gaithersburg, MD................ (301) 948 5717
CONNECTION-80 Gaithersburg, MD .... (301) 840 8588 *24
Hex Silver Spring, MD.............................. % (301) 593 7033 *24
NET-WORKS COMM Center NW3NAGAD, Laurel, MD . (301) 953 1110
PMS—Baltimore, MD................................ (301) 764 1995 *24
PMS—Ellicott City, MD .... (301) 465 3176
PMS—Pikesville, MD .... (301) 653 3413
PSBBS Baltimore, MD............................... (301) 994 0399 *24
RCP/M RBBS Laurel, MD ....! (301) 953 3753 *24
RCP/M RBBS Bethesda, MD ....! (301) 229 3196
Remote Northstar NASA, Greenbelt, MD .... (301) 344 9156

303

ABBS Denver, CO .... (303) 759 2625
CONNECTION-80 Denver, CO .... (303) 690 4566 *24
CoxCo, Arvada, CO .... (303) 423 5001 *24
FORUM-80 #1, Denver, CO .... (303) 341 0636 *24
FORUM-80 #2, Denver, CO .... (303) 399 8858 *24
HBBS Denver, CO .... (303) 343 8401 *24
RCP/M CUG-NOTE, Denver, CO.....................! (303) 781 4937 *24
RCP/RBBS Boulder, CO .... (303) 499 9169
RCP/RBBS Arvada Elect., Colorado Springs, CO. .... (303) 634 1158 *24
Remote Northstar Denver, CO .... (303) 444 7231

305

ABBS Byte Shop, Ft. Lauderdale, FL .... (305) 486 2983
ABBS Byte Shop, Miami, FL.......................... (305) 261 3639
ABBS West Palm Beach, FL .... (305) 848 3802
AMIS APOGEE Miami, FL .... (305) 238-1231 -rb
BBS Homestead, FL................................. (305) 246 1111
CONNECTION-80 Orlando, FL .... (305) 644 8327 *24
CONNECTION-80 Winter Garden, FL .... (305) 877 2829 *24
Corsair, Goldcoast, FL .... (305) 968 8653
FORUM-80 Ft. Lauderdale, FL .... (305) 772 4444 *24
Greene Machine, WPB, FL .... (305) 965 4388 -so
INFOEX-80 West Palm Beach, FL .... (305) 683 6044 *24
MOUSE-NET Orlando, FL............................ (305) 277 0473 *24
Micro-80 West Palm Beach, FL....................... (305) 686 3695
NET-WORKS Big Apple, Miami, FL .... (305) 948 8000

Personal Message System-80, Deerfield Bch, FL . . . . (305) 427 6300 *24
RCP/M RBBS South Florida . . . . (305) 255 6027
TRADE-80 Ft. Lauderdale, FL . . . .#1 (305) 525 1192

309

ABBS Peoria, IL. . . . . (309) 692 6502
NET-WORKS MAGIE, Galesburg, IL. . . . . (309) 342 7178
RCP/M Geneseo, IL . . . . (309) 944 5455

312

ABBS AIMS, Oak Brook, IL. . . . . (312) 789 0499 *24
ABBS CODE, Glen Ellyn, IL . . . . (312) 537 7063 *24
ABBS Downers Grove, IL . . . . (312) 964 7768
ABBS Gamemaster, Chicago, IL . . . . (312) 475 4884 *24
ABBS Illini Microcomputer, Naperville, IL . . . . (312) 420 7995
ABBS Oak Brook Computer, Oak Brook, IL. . . . . (312) 941 9009
ABBS Rogers Park, Chicago, IL . . . . (312) 973 2227
AMIS Chicago, IL . . . . (312) 789 3610 *24
BBS Electronic Exchange, Chicago, IL. . . . . (312) 541 6470 *24
CBBS Chicago, IL . . . .#1 (312) 545 8086 *24
FBBS Skokie, IL. . . . .#1 (312) 677 8514
MARS/RP Rogers Park, IL . . . . (312) 743 8176 *24
MCMS C.A.M.S. Chicago, IL . . . .#1 (312) 927 1020 *24
MCMS L.A.M.S. Round Lake, IL . . . . (312) 740 9128
MCMS Message-82, Chicago, IL . . . . (312) 622 4442 *24
MCMS Metro West Database, Chicago, IL . . . . (312) 260 0640 *24
MCMS WACO Hot Line, Schaumburg, IL. . .(pvt) . . . . (312) 351 4374 *24
NESSY Chicago, IL . . . .#1 (312) 773 3308
NESSY Terry and Gwen's, Palatine, IL . . . . (312) 289 6393
NET-WORKS CLAH, Arlington, Heights, IL . . . . (312) 255 6489
NET-WORKS Pirate's Ship, Chicago, IL. . . . . (312) 935 2933 *24
PBBS Co-operative Comp Svc, Palatine, IL. . . . . (312) 359 9450 *24
PMS - Arlington Heights, IL . . . . (312) 870 7176 *24
PMS - Chicago, IL . . . . (312) 373 8057 *24
PMS - I.A.C., Lake Forest, IL. . . . . (312) 295 6926 *24
RATS Homewood, IL . . . . (312) 957 3924
RCP/M AIMS Hinsdale, IL . . . .! (312) 789 0499
RCP/M CBBS HUG, Chicago, IL . . . . (312) 671 4992 *24
RCP/M IBM-PC, Niles, IL. . . . .&! (312) 647 7636 *24
RCP/M Logan Square, Chicago, IL . . . . (312) 252 2136
RCP/M NEI, Chicago, IL . . . . (312) 949 6189
RCP/M Palatine, IL . . . .& (312) 359 2553 *24
RCP/M RBBS Hyde Park, IL . . . . (312) 955 4493
RCP/M TRS-80, Chicago, IL . . . . (312) 949 6189

313

ABBS Detroit, MI	(313) 477 4471
ABBS Michigan Apple-Fone, Southfield, MI	(313) 357 1422
AMIS A.R.C.A.D.E. Sterling Heights, MI	(313) 978 8087 *24
AMIS M.A.C.E. Detroit, MI ....#1	(313) 868 2064 *24
BULLET-80 Mt. Clemens, MI	(313) 465 9531
FORUM-80 Pontiac, MI	(313) 335 8456
RCP/M Detroit, MI................................	(313) 584 1044 -rb
RCP/M MCBBS Keith Petersen, Royal Oak, MI	(313) 759 6569 -rb
RCP/M MCBBS TCBBS Dearborn, MI	(313) 846 6127 *24
RCP/M RBBS Southfield, MI........................	(313) 559 5326 *24
RCP/M RBBS Westland, MI	(313) 729 1905 -rb
Westside Download, Detroit, MI	(313) 533 0254

314

ABBS Century Next Computers, St. Louis, MO	(314) 442 6502
ABBS St. Louis, MO	(314) 838 7784 *24
Midwest, St. Louis, MO............................	(314) 227 4312 -so
NET-WORKS Computer Station, St. Louis, MO	(314) 432 7120
NET-WORKS St. Louis, MO	(314) 781 1308
PET BBS Commodore Comm., Lake St. Louis, MO ......!	(314) 625 4576 *24
SLAMS St. Louis, MO.............................	(314) 839 4307

316

FORUM-80, Wichita, KA..........................&	(316) 682 2113 *24

317

NET-WORKS Greenfield, IN	(317) 326 3833 *24
ONLINE Indianapolis, IN (ID# = GUES, pswd = pass)	(317) 787 9881 *24
PMS - Indianapolis, IN	(317) 787 5486 *24

318

FORUM-80, Shreveport, LA.........................	(318) 631 7107 *24

319

ABBS Apple-Med, Iowa City, IA	(319) 353 6528
CBBS Cedar Rapids, IA............................	(319) 364 0811

401

NET-WORKS Computer City, Providence, RI . . . .   (401) 331 8450 *24

402

ABBS Omaha, NE . . . .   (402) 339 7809
TRADE-80, Omaha, NE . . . .   (402) 292 6184

403

RCP/M RBBS Computron, Edmonton, Alberta, Can. . . . .   (403) 482 6854 *24
RCP/M RBBS Edmonton, Alberta, Canada . . . .   (403) 454 6093 *24

404

ABBS #X, Atlanta, GA . . . .   (404) 256 1549
ABBS AGS, Atlanta, GA . . . .   (404) 733 3461 *24
ABBS Baileys Computer Store, Augusta, GA . . . .   (404) 790 8614
BULLET-80 Fayetteville, GA . . . .   (404) 461 9686
CBBS Atlanta, GA . . . .   (404) 394 4220 *24
Remote Northstar Atlanta, GA . . . . #1 (404) 926 4318 *24
Scanboard-80, Atlanta, GA . . . .   (404) 457 8384
Telemessage-80, Atlanta, GA . . . .   (404) 962 0616

405

HBBS Oklahoma City, OK . . . .   (405) 848 9329 *24

406

BBS IBM PC Billings, MT . . . .   (406) 656 9624

408

AMIS GRAFEX Cupertino, CA . . . .   (408) 253 5216
AMIS IBBBS San Jose, CA . . . .   (408) 298 6930
AMIS T.A.B.B.S. Sunnyvale, CA . . . .   (408) 942 6975
BULLET-80 San Jose, CA . . . .   (408) 241 0769
NET-WORKS Computer Emporium, San Jose, CA . . . .   (408) 227 0227
PMS - Campbell, CA . . . .   (408) 370 0873 *24
PMS - Santa Clara, CA . . . .   (408) 554 9036
PMS - Santa Cruz, Aptos, CA . . . .   (408) 688 9629 *24

RCP/M Collossal Oxgate, ?? ....! (408) 263 2588
RCP/M RBBS San Jose Oxgate, San Jose, CA ....! (408) 287 5901 *24
RCP/M Silicon Valley, CA ....! (408) 246 5014 *24

412

CBBS PACC, Pittsburgh, PA .... (412) 822 7176 *24

414

ABBS Colortron Computer, WI .... (414) 637 9990 *24
BBS SUE Milwaukee, WI .... (414) 483 4578
CBBS MAUDE Milwaukee, WI .... (414) 241 8364 *24
PET BBS S.E.W.P.U.G., Racine, WI .... (414) 554 9520 *24
RBBS Big Top, Milwaukee, WI .... (414) 259 9475

415

ABBS Computerland, Fremont, CA.................... (415) 794 9314
ABBS Hayward, CA .... (415) 881 5662
ABBS PCnet, San Francisco, CA .... (415) 863 4703 *24
ABBS South of Market, San Francisco, CA .... (415) 469 8111 -so
BBS Living Videotext, Menlo Park, CA .... (415) 327 8876 *24
BULLET-80 Redwood City, CA .... (415) 367 1339 *24
CBBS Lambda, Berkeley, CA .... (415) 658 2919 -so
CBBS Proxima, Berkeley, CA .... (415) 357 1130
CONFERENCE-TREE #2, San Francisco, CA .... (415) 928 0641
CONFERENCE-TREE #3, Hayward, CA................ (415) 538 3580
CONNECTION-80 Fremont, CA...................... (415) 651 4147 *24
Databoard BBS, Atherton, CA ....! (415) 367 7638 *24
HMS Horny Message System, Oakland, CA .... (415) 845 2079 -so
IAC Message Base, Menlo Park, CA.................... (415) 367 1339 *24
INFO-NET Foster City, CA .... (415) 349 3126
Kinky Kumputer, San Francisco, CA .... (415) 647 9524 -so
PMS - Pleasanton, CA .... (415) 462 7419 *24
PMS - Portola, Valley, CA .... (415) 851 3453 *24
RCP/M Piconet Oxgate, ?? .... (415) 965 4097 *24
RCP/M RBBS DataTech, San Carlos, CA .... (415) 595 0541
RCP/M RBBS Larkspur, CA .... (415) 461 7726 *24
RCP/M RBBS Marin County, CA ....! (415) 383 0473 *24
Sunrise Omega-80, Oakland, CA .... (415) 452 0350

416

RCP/M HAPN Hamilton, Ontario, CN ....! (416) 335 6620 *24
RCP/M Mississauga HUG, Toronto, Ontario, CN .....$& (416) 826 5394 *24

419

ABBS ABACUS II, Toledo, OH . . . . (419) 865 1594
ABBS Computer Store, Toledo, OH . . . . (419) 531 3845

501

CONNECTION-80 Little Rock, AS . . . . (501) 372 0576
PMS - Ft. Smith Computer Club, Ft. Smith, AK . . . . ! (501) 646 0197

502

ABBS Louisville, KY . . . . (502) 245 7811 *24

503

CBBS NW, Portland, OR . . . . (503) 646 5510 *24
CONNECTION-80 PAUG, Portland, OR . . . . (503) 281 7653
FORUM-80 Medford, OR . . . . (503) 535 6883 *24
OARCS Portland, Oregon . . . . (503) 641 2798
RCP/M Chuck Forsberg, OR . . . . ! (503) 621 3193

504

ABBS Baton Rouge, LA . . . . (504) 291 1360
Baton Rouge Data System, Baton Rouge, LA . . . . (504) 926 0181
CBBS Baton Rouge, LA . . . . (504) 273 3116 *24
NET-WORKS Crescent City, Baton Rouge, LA . . . . (504) 454 6688

509

ABBS Electro-Mart, Spokane, WA . . . . (509) 534 2419 *24
ABBS Rob Roy Computer, Yakima, WA . . . . (509) 575 7704
ABBS Spokane, WA . . . . (509) 456 8900
PMS - Richland, WA . . . . (509) 943 6502 *24

512

ABBS San Antonio, TX . . . . (512) 737 0214 *24
Bathroom Wall BBS, San Antonio, TX . . . . ! (512) 655 8143
CBBS Corpus Christi, TX . . . . (512) 855 1512
CONFERENCE-THREE Victoria, TX . . . . (512) 578 5833
FORUM-80 San Antonio, TX . . . . (512) 340 6720

NET-WORKS Sparklin' City, Corpus Christi, TX . . . . (512) 882 6569
ORACLE Classified System, Austin, TX . . . . (512) 346 4495
SATUG BBS, San Antonio, TX . . . . ! (512) 494 0285

513

BBS-80 Cincinnati, OH . . . . (513) 244 2983
NET-WORKS Dayton, OH . . . . (513) 223 3672
PMS - Cincinnati, OH . . . . (513) 671 2753
XBBS Hamilton, OH . . . . (513) 863 7681 *24

514

NET-WORKS Montreal, Canada . . . . (514) 937 2188 *24

515

NET-WORKS Computer Emporium, Des Moines, IA . . . . . (515) 279 8863

516

ABBS Pirates Cove, Long Island, NY . . . . (516) 698 4008
Adventure BBS . . . . (516) 621 9296
CBBS LICA LIMBS, Long Island, NY . . . . (516) 561 6590 *24
CBBS Long Island, NY . . . . (516) 334 3134 *24
CONNECTION-80 Centereach, NY . . . . (516) 588 5836
CONNECTION-80 Great Neck, NY . . . . (516) 482 8491 *24
RCP/M New York, NY . . . . (516) 791 1767
RCP/M Long Island, NY . . . . (516) 698 8619 -rb

517

CONNECTION-80 Lansing, MI . . . . (517) 339 3367

518

Capital City BBS, Albany, NY . . . . (518) 346 3596 *24
FORUM-80 Albany, NY . . . . (518) 785 8478

601

Remote Apple Jackson, MS . . . . (601) 992 1918 *24

602

A-C-C-E-S-S Phoenix, AZ	..........................&	(602) 957 4428 *24
A-C-C-E-S-S Phoenix, AZ	..........................#1	(602) 996 9709 *24
A-C-C-E-S-S Phoenix, AZ	....	(602) 274 5964
A-C-C-E-S-S Scotsdale, AZ	....	(602) 998 9411 *24
ABBS Phoenix, AZ	....	(602) 898 0891
CBBS TSG, Tucson, AZ	....	(602) 746 3956 *24
FORUM-80 Sierra Vista, AZ	....	(602) 458 3850 *24

603

ABBS VIBBS, Nashua, NH	....	(603) 888 6648
CONNECTION-80 Peterborough, NH	....	(603) 924 7920
FORUM-80, Nashua, NH	....	(603) 882 5041
MOUSE-NET Nashua, NH	....	(603) 673 9476 *24
NET-WORKS Portsmouth, NH	....	(603) 436 3461

604

ABBS Vancouver, B.C.	....	(604) 437 7001
CBBS Prince George, B.C., Canada	....	(604) 562 9515
CBBS Vancouver, BC, Canada	....	(604) 687 2640 *24
RCP/M CBBS Frog Hollow, Vancouver, BC, CN	.......!	(604) 873 4007 *24
RCP/M CBBS Vancouver, BC, Canada	....	(604) 687 2640 *24
RCP/M RBBS Surrey, Vancouver, BC, CN	.............!	(604) 584 2643 *24
RCP/M Terry O'Brien, Vancouver, BC, Canada	....	(604) 584 2543

607

RCP/M SJBBS Johnson City, NY	....	(607) 797 6416

608

AMIS Magic Lantern, Madison, WI	....	(608) 251 8538

609

ABBS Turnersville, NJ	....	(609) 228 1149
RATS Wenonah, NJ	....	(609) 468 5293
RATS Wenonah #2	....	(609) 468 3844

612

ABBS Calvary Mission Church, Minneapolis, MN ....! (612) 471 0252
CBBS Richfield, MN .... (612) 869 5780
CONFERENCE-TREE Minneapolis, MN............... (612) 227 0307
MCMS NC Software, Minneapolis, MN .... (612) 533 1957 *24
PMS - Minneapolis, MN .... (612) 929 6699 *24
ZBBS Minneapolis, MN............................. (612) 426 9028

613

ABBS Compumart, Ottawa, Ontario, Canada .... (613) 725 2243

614

BULLET-80 Ironton, OH .... (614) 532 6920
RCP/M CBBS Columbus, OH .... (614) 272 2227 *24

616

ABBS Computer Room, Kalamazoo, MI .... (616) 382 0101
AMIS G.R.A.S.S. Grand Rapids, MI.................. (616) 241 1971 *24
CONNECTION-80 W. Mich. Micro Group, MI .... (616) 457 1840 *24

617

AMIS Starbase 12 Philadelphia, PA .... (617) 876 4885
BULLET-80 Boston, MA ....& (617) 266 7789 *24
Boston Information Exchange, Boston, MA ....& (617) 423 6985 *24
CBBS Boston, MA .... (617) 646 3610 *24
CBBS Lawrence General Hospital, Boston, MA .... (617) 683 2119
FORUM-80 Westford, MA .... (617) 692 3973
NET-WORKS Pirate's Harbor, Boston, MA .... (617) 738 5051
NMS Natick, MA .... (617) 653 4282 *24
New England Comp. Soc., Maynard, MA .... (617) 897 0346
PMS - Apple Guild, Weymouth, MA .... (617) 767 1303 *24
PMS - Computer City, Danvers, MA .... (617) 774 7516
RCP/M MCBBS Superbrain, Lexington, MA ....$& (617) 862 0781 *24

618

NET-WORKS Granite City, IL .... (618) 877 2904
NET-WORKS Warlock's Castle St. Louis, MO .... (618) 345 6638

701

NET-WORKS Armadillo, Grand Fork, ND . . . .   (701) 746 4959

702

ABBS Apple Cider, Las Vegas, NV . . . .   (702) 454 3417
FORUM-80 Las Vegas, NV . . . .   (702) 362 3609 *24
PMS - Century 23, Las Vegas, NV . . . . !   (702) 878 9106 *24
SIGNON Reno, NV . . . . (pswd = FREE) . . . . .   (702) 826 7234

703

ABBS Software Sorcery, Herndon, VA . . . . !   (703) 471 0610
BBS Annandale, VA . . . .   (703) 978 9754
BBS IBM PCUG Annandale, VA . . . .   (703) 560 0979
C-HUG Bulletin Board, Fairfax, VA . . . .   (703) 360 3812 *24
CBBS AMRAD, Washington, DC . . . .   (703) 734 1387 *24
Carrier 2 Alexandria, VA . . . .   (703) 823 5210
FORUM-80 Prince William County, VA . . . .   (703) 670 5881 *24
MICRO-COM ? . . . .   (703) 780 0610
Potomac Micro Magic Inc., Falls Church, VA . . . .   (703) 379 0303 *24
RCP/M Arlington, VA . . . . !   (703) 536 3769
RCP/M CBBS RLP, MacLean, VA . . . . !   (703) 524 2549 *24
TCUG BBS, Washington, DC . . . .   (703) 451 8475 *24

704

ABBS Charlotte, NC . . . . !   (704) 364 5254

707

RCP/M RBBS Napa Valley, CA . . . .   (707) 253 1523

713

ABBS Madam Bokeatha Society, Houston, TX . . . .   (713) 455 9502
ABBS Youngs Elect Svc., College Station, TX . . . .   (713) 693 3462 *24
BULLET-80 Houston, TX . . . .   (713) 331 2599
ONLINE Houston, TX . . . .   (713) 528 7025 *24
PMS - Gulfcoast, Freeport, TX . . . .   (713) 233 7943 *24

714

ABBS Newport Beach, CA............................!	(714) 645 5256 *24
BULLET-80 Orange County, Anaheim, CA	(714) 952 2110
BULLET-80 Riverside, CA	(714) 359 3189
COMM-80 OCTUG..................................	(714) 530 8226
CONNECTION-80 Escondido, CA ....!	(714) 746 6265
DIAL-YOUR-MATCH #11.........................!	(714) 242 1882
Dimension-80 Orange, CA	(714) 974 9788
FORUM-80 Orange County, Anaheim, CA	(714) 545 9549
NET-WORKS Info-Net, Costa Mesa, CA	(714) 545 7359
North Orange County Computer Club	(714) 633 5240
ONLINE CDC, San Diego, CA ....!	(714) 452 6012
ONLINE Saba, San Diego, CA ....!	(714) 291 5229
ONLINE Santee, CA .(ID# = GUEST, pswd = PASS).....	(714) 561 7271 *24
Orange County Data Exchange, Garden Grove, CA	(714) 537 7913
PMS - **IF**, Anaheim, CA.........................	(714) 772 8868 *24
PMS - Escondido, CA	(714) 746 0667
PMS - Lakeside, CA (type PMS to activate)	(714) 561 7271 *24
PMS - Mesa Systems, San Diego, CA ....!	(714) 271 8613
PMS - Mission Valley, CA	(714) 295 8280
PMS - San Diego, CA...............................	(714) 582 9557 *24
PMS - Santee, CA.................................#1	(714) 561 7277 *24
RCP/M CBBS ANAHUG, Anaheim, CA	(714) 774 7860
RCP/M RBBS San Diego, CA ....$&!	(714) 273 4354 *24
RCP/M RBS GFRN Dta Exch., Garden Grove, CA ....$&	(714) 534 1547 *24
RCP/M RBBS ANAHUG, Anaheim, CA	(714) 774 7860 *24

716

CBBS RAMS, Rochester, NY	(716) 244 9531
RCP/M RBBS Rochester, NY	(716) 223 1100 *24

717

BULLET-80 Clarks Summit, PA	(717) 586 2112

802

ABBS Vermont, Essex Junction, VT...................	(802) 879 4981 *24
ST80-CC Lance Micklus, Inc., Burlington, VT ....#1	(802) 862 7023 *24

803

Compusystems, Columbia, SC .... (803) 771 0922
FORUM-80 Charleston, SC .... (803) 552 1612 *24
FORUM-80 Augusta, GA .... (803) 279 5392
RCP/M RBBS Fort Mill, SC........................ (803) 548 0900 *24

804

NBBS Norfolk, VA................................! (804) 444 3392
RCP/M RBBS Grafton, VA ....! (804) 898 7493
Remote Northstar Virginia Beach, VA .... (804) 340 5246
Talk-80 ROBB, Portsmouth, VA .... (804) 484 9636

805

BULLET-80 Lancaster, CA .... (805) 947 9925
Computer Arts Message System, Newhall, CA .... (805) 255 6445
RCP/M CBBS CP/M Net Simi Valley, CA .... (805) 527 9321
RCP/M RBBS Thousand Oaks, CA.................... (805) 496 9522 *24
Remote Northstar Santa Barbara, CA .... (805) 682 7876
Remote Northstar Santa Barbara, CA .... (805) 964 4115

806

ABBS Computer Corner, Amarillo, TX .... (806) 355 5610
BULLET-80 Littlefield, TX.......................... (806) 385 6843

808

NET-WORKS Hawaii .... (808) 521 7312

812

CBBS Bloomington, IN .... (812) 334 2522

813

Alpha, Tampa, FL . . . (acct# = ABCDOO, pwd = TRYIT) .. (813) 251 4095 *24
Apollo's Chariot, Apollo, FL .... (813) 645 3669
BSBB Tampa, FL .... (813) 885 6187
Bradley Computer BBS .... (813) 734 7103
CONNECTION-80 Tampa, FL .... (813) 977 0989

Micro Informer. . . . .   (813) 884 1506
North Largo, FL. . . . .   (813) 535 9341 *24
RCP/M RBBS Computerized Services, Tampa, FL . . . .   (813) 839 6746
Remote Northstar Largo, FL. . . . .   (813) 535 9341 *24
Wild goose board, ? . . . .   (813) 988 7400

814

RCP/M CUG-NODE, PA State College . . . . ! (814) 238 4857 *24
TRADE-80 Erie, PA . . . .   (814) 898 2952 *24

815

MCMS J.A.M.S. Lockport, IL. . . . .   (815) 838 1020 *24

816

FORUM-80 Kansas City, MO. . . . .& (816) 931 9316
FORUM-80 Kansas City, MO . . . .#1 & (816) 861 7040 *24
Market 80, Kansas City, MO . . . .   (816) 931 9316

817

ABBS Teledunjon I, Dallas, TX . . . .   (817) 469 1626
CBBS Waco, TX . . . .   (817) 776 1375
FORUM-80 Wichita Falls, TX . . . .   (817) 855 3916
NET-WORKS Computer Pro, Ft. Worth, TX . . . .   (817) 732 1787

901

ABBS Computer Lab, Memphis, TN . . . .   (901) 761 4743
ABBS Memphis, TN . . . .   (901) 725 5691
FORUM-80 Medical, Memphis, TN . . . .   (901) 276 8196 *24

904

ABBS Fort Walton Beach, Destin, FL . . . .   (904) 243 1257
BBS Pensacola, FL. . . . .   (904) 477 8783

907

ABBS Ketchikan, AK. . . . .   (907) 225 6789

CONFERENCE-THREE Anchorage, AK . . . . (907) 344 5251
PMS - Anchorage, AK . . . . (907) 344 8558

912

DIAL-YOUR-MATCH #3 . . . . (912) 233 0863 -so

913

Experimental-80 Kansas City, MO . . . . (913) 676 3613
ONLINE Dickinsons Movie Guide, Mission, KS . . . . (913) 432 5544
ONLINE ON-LINE Computer Center, KC, MO . . . . (913) 341 7987
PMS - Apple Bits, Kansas City, MO . . . . (913) 341 3502 *24
PMS - Your Computer Connection, KS Cty, MO . . . . (913) 381 1021
RCP/M RBBS MUG, Mission, KS . . . .& (913) 362 9583 *24

914

BULLET-80 Poughkeepsie, NY . . . . (914) 278 2375 -so
Nybbles-80 Elmsford, NY . . . . (914) 592 5385
RCP/M RBBS Woodstock, NY . . . .! (914) 679 8734
RCP/M SJBBS Bearsville, NY . . . . (914) 679 6559 -rb
ST80-PBB Monroe Camera Shop, Monroe, NY . . . . (914) 782 7605

915

FORUM-80 El Paso, TX . . . . (915) 755 1000 *24
HBBS El Paso, TX . . . . (915) 592 1910
RCP/M RBBS Comp. Tech. Assoc., El Paso, TX . . . .! (915) 533 2202 *24
RCP/M RBBS El Paso, TX. . . . . (915) 598 1668

916

Aviators Bulletin Board, Sacramento, CA . . . . ,. . . . (916) 393 4559
RCP/M CBBS Sacramento, CA . . . .! (916) 483 8718 *24

918

BULLET-80 Tulsa, OK . . . . (918) 749 0059 *24
CONNECTION-80 Tulsa, OK . . . . (918) 747 1310 *24

Help Files of Popular System Types

The following files were all downloaded from PAMS that each seemed representative of their type (their formats as to line length, capitalization are just as you would read them off your screen). Most of the commands listed here for a certain system should work on nearly all other systems of that type.

ABBS Systems

The ABBS letters stand for Apple Bulletin-Board System. The software runs exclusively on the Apple II although any type of computer may call one of these systems. The help file is as follows:

Function:

(A,B,C,D,E,G,H,K,L,N,Q,R,S,T,V,W,X,
DOWNLOAD,UPLOAD,NEWS,CONFERENCES,MAGUS,?)?**H**

Enter your Choice:

'ALL' for complete review
'CTRL' for CTRL Characters, or
'A,B,C,D,E,G,H,K,L,N,Q,R,S,T,V,
W,X,?'
CONF(erences)
NEWS
DOWNLOAD
UPLOAD
For Individual Response.

   C/R   to return to ABBS

?**ALL**

FOLLOWING IS A BRIEF LIST AND DESCRIPTION OF THE COMMANDS AND
THEIR USAGE:

CTRL C—CANCELS PRINTING OF THE CURRENT LINE BEING PRINTED.

CTRL H (BACKSPACE)—ALLOWS YOU TO BACKSPACE ONE CHARACTER AT
   A TIME AND PRINTS A ' ' FOLLOWED BY THE CHARACTER YOU ARE
   BACKSPACING OVER. THIS IS THE SAME ROUTINE AS IS USED FOR
   DELETE OR RUBOUT INSTEAD OF TRUE DELETE. (FOR THE BENEFIT OF
   PRINTERS)

CTRL K—CANCELS CURRENT FUNCTION

CTRL U (FORWARD ARROW)—STARTS YOU BACK AT THE BEGINNING OF
   THE CURRENT LINE BEING TYPED. (I.E. START OVER)

A—TOGGLE DISPLAY MODE BETWEEN 64 and 40 COLUMN MODE. WHEN YOU SIGN ON YOU ARE ALLOWED TO INPUT UP TO 64 CHARS/LINE.

A BELL WILL SOUND AT 59 AND ON UP TO 64 COLUMNS AT WHICH POINT YOU WOULD BE FORCED ONTO THE NEXT LINE OF TEXT.

IN THE APPLE 40 MODE, THE BELL WILL RING AT 35, THEN AGAIN AT 38 AND 39. DROPPING YOU TO THE NEXT LINE AT 39. 39 WAS USED INSTEAD OF 40 TO AVOID AN EXTRA BLANK LINE BECAUSE OF THE 40TH CHARACTER.

B—PRINT BULLETIN. REPRINTS BULLETINS THAT YOU SEE WHEN YOU SIGN ON.

C—CASE SWITCH. WHEN YOU SIGN ON THE SYSTEM DEFAULTS TO UPPER CASE INPUT & OUTPUT. THE C COMMAND ALLOWS YOU TO SWITCH BE-TWEEN UPPER CASE ONLY MODE AND UPPER/LOWER CASE MODE.

D—DUPLEX SWITCH. ALTERNATELY SELECTS FULL OR HALF DUPLEX OPERATION AND INFORMS YOU OF CURRENT STATUS.

E—ENTER MESSAGE. ALLOWS YOU TO ENTER A MESSAGE INTO SYSTEM. ENTRY COMMANDS ARE BASICALLY SELF EXPLANATORY. A CARRIAGE RETURN (C/R) AT THIS POINT WILL LIST OUT THE COMMAND MENU FOR ENTRIES. THE CHANGE COMMAND ALLOWS YOU TO CHANGE AN EN-TIRE LINE OR JUST PART OF IT. TO CHANGE JUST PART OF A LINE ENTER THE FOLLOWING: /STRING TO CHANGE/REPLACEMENT STRING/ YOU WILL THEN BE SHOWN THE NEW VERSION OF THE LINE AND SHOULD HIT RETURN TO LEAVE THE CHANGE MODE. YOU CAN EXIT THE ENTER MODE ENTIRELY BY ENTERING AN A COMMAND. YOU SHOULD BE SURE TO ENTER AN S TO SAVE A MESSAGE TO DISC.

G—GOODBYE. EXITS PROGRAM & HANGS UP PHONE ON ABBS END OF CONNECTION.
C/R   TO CONTINUE,   E   TO END?

H—HELP. PRINTS THIS ROUTINE.

K—KILL A MESSAGE. ENTER THIS TO DELETE A MESSAGE FROM THE FILE. A PASSWORD MAY BE NECESSARY IF ONE WAS USED AT THE TIME OF MESSAGE ENTRY.

L—LINE FEED ON/OFF. NORMALLY ON. FOR TERMINALS THAT NEED AN EXTRA LINE-FEED CHARACTER TO ADVANCE TO THE NEXT LINE.

N—NULLS. ADDS AN EXTRA DELAY AFTER A CARRIAGE RETURN TO ALLOW PRINTERS TIME TO MOVE THE PRINTERHEAD BACK TO STARTING POSITION. THIS OPTION ONLY WORKS WITH THE LINE FEED OPTION ON. EACH NULL IS EQUIVALENT TO 30 MILLISECONDS DELAY AND IS ADJUST-

ABLE FROM 1 TO 30. IT DEFAULTS TO ONE.

Q—QUICK SCAN. AN ABBREVIATED SCAN. SEE 'S'

R—RETRIEVE MESSAGES. ALLOWS YOU TO RETRIEVE A MESSAGE FROM THE FILE. YOU CAN SPECIFY EITHER TO RETRIEVE A SINGLE MESSAGE OR MULTIPLE MESSAGES OR ALL THE MESSAGES FROM A GIVEN NUMBER ON UP. TO SPECIFY A SINGLE MESSAGE, ENTER R;N WHERE N IS THE # OF THE MESSAGE YOU WISH TO READ. TO READ MULTIPLE MESSAGES SEPARATE EACH MSG. # WITH A ';'. (IE. R;1;2;3;7;8) TO RETRIEVE ALL THE MESSAGES STARTING WITH A CERTAIN MSG. # ENTER AN R FOLLOWED BY A ';' FOLLOWED BY THE STARTING NUMBER, FOLLOWED BY AN UP-ARROW. (IE. R;10;   )

S—SCAN MESSAGES. ALLOWS YOU TO SCAN OVER MESSAGES STARTING AT THE MESSAGE NUMBER THAT YOU SPECIFY.

T—TIME AND DATE. GIVES YOU THE CURRENT TIME AND DATE. THIS IS ALSO USED AUTOMATICALLY DURING LOG-IN.

W—WELCOME. PRINTS WELCOME MESSAGE AT BEGINNING OF PROGRAM.

X—EXPERT USER. DOES AWAY WITH CERTAIN EXPLANATORY MESSAGES DURING THE PROGRAM. IT ALSO ALLOWS CERTAIN C/R DEFAULTS. EX: A C/R IN RESPONSE TO FUNCTIONS? WILL PRINT FUNCTIONS SUPPORTED BY THE SYSTEM.

—PRINTS FUNCTIONS SUPPORTED IN THAT CURRENT MODE OF OPERATION.

CONF—ALLOWS YOU TO SWITCH BETWEEN SETS OF MESSAGES. (ONLY VALID IF THE OPTIONAL CONFERENCE MODULE IS INSTALLED)

NEWS—DISPLAYS A LIST OF FILES THAT ARE AVAILABLE FOR READING. (VALID ONLY IF OPTIONAL NEWS MODULE IS INSTALLED)

DOWNLOAD—GIVES YOU ACCESS TO A LIBRARY OF SOFTWARE THAT CAN BE DOWNLOADED TO YOUR SYSTEM. (ONLY VALID IF THE OPTIONAL DOWNLOAD MODULE IS INSTALLED)

UPLOAD—ALLOWS YOU TO TRANSFER A FILE TO THE SYSOP. (ONLY VALID IF THE OPTIONAL UPLOAD MODULE IS INSTALLED)

Bullet-80 Systems

The Bullet-80 software runs on TRS-80 machines. But, people with all types of computers do call in.

These are the commands:

[A]rcade	[B]ulletins
[C]onference	[D]ownload
[E]lectromart	[H]ow long
[I]nstructions	[M]essage Base
[O]ther Numbers	[P]rogram Uploading
[S]ystem Config	[T]erminate
[U]serlog	[X]pert user

What is your choice? **Instructions**

System Instructions

Press [S] to Stop or [P] to pause

To use this system is simple, just press the FIRST letter of any command you care to execute, entering a RETURN is not needed because Bullet-80 makes extensive use of INKEY$. Entering a Carriage Return, [C/R], or pressing [ENTER] at an abbreviated menu will display an expanded if you're not sure what to do.

Here is an explanation of the commands:

[B ulletins]— This command will display the current system bulletins.

[C]hat   — This will allow you to chat with the sysop.

[D]ownload — This module allows you to get programs from database.

[E]lectronic
Shopping   — This module opens up for you a new use for your home or business computer. It offers you a selection of items that you could purchase and bill to your charge card. It's easy and fun!

[I]nstruction — Will print out this list.

[L]eave
Message   — Allows you to leave messages for others to read. The message size is limited to 16 lines and 64 characters. Messages can be OPEN or PRIVATE.

[O]ther System
Phone Numbers   — Prints out a list of many on-line systems.

[P]rogram
Uploading   — This module will allow you to upload programs to Bullet-80 so others can enjoy your ORIGINAL or PUBLIC DOMAIN programs. After review they will be placed in our downloading section.

[R]etrieve   — This will let you read all messages that are available to you. You have the option of a [F]orward retrieval, [R]everse retrieval, a [S]ingle retrieval or [N]ew retrieval. During the retrieving of messages you can press [S], [P], or [N] for: Stopping, Pausing, or going to the Next message.

[S]can   — allows you to get a quick scan of messages by displaying only an abbreviated header. The options of [S] or [P] are active here.

[T]erminate   — Logs you off of Bullet-80. At this time you can leave a private message to the sysop. Always use this option so your information can be recorded.

[U]sers Log   — Will display Bullet-80 users for you. You can scan [F]orward or [R]everse or even [S]earch for a certain user.

[V]iew System
Configuration— Will tell you what the hardware for Bullet-80 is made up of.

CBBS Systems

The letters stand for Computer Bulletin Board System and these were among the first such systems running. Software was done by telecommunicating pioneer Ward Christiansen.

FOUR CONTROL KEYS ARE DEFINED (THEY CAN BE CHANGED, TYPE N).
   CNTL   A ABORTS TO THE EXECUTIVE,   CNTL   H RUBS OUT ONE CHAR.   CTRL   S FREEZES THE PRINTOUT,   CNTL   Q RESUMES PRINTING. FOR MORE DETAILS ON HOW TO USE THIS TYPE H FOR HELP.

TO ANY NEWCOMERS, PASSWORDS ARE ONLY REQUIRED TO EDIT OR CREATE A MESSAGE. THEY ARE MADE UP BY THE CREATOR OF THE MESSAGE.

EXEC COMMAND (P,L,E,S,C,D,N,X,M,U,H = HELP)**H**
CBBS EXEC LEVEL COMMANDS ARE INVOKED BY 1 KEY
EACH MESSAGE IS UP TO 16 LINES OF UP TO 64 CHARS
THE FIRST LINE OF A MESSAGE IS CALLED THE TITLE
P — LIST EACH TITLE LINE, THEN PAUSE
L — LIST EACH TITLE
E — EDIT A MESSAGE
S — TYPE OUT CBBS STATUS
C — CREATE A NEW MESSAGE

D — DELETE A MESSAGE
N — CHANGE CONTROL CHARS AND OTHER PARAMETERS
X — SIGN OFF
M — TYPE OUT A MESSAGE
U — DECLARE YOURSELF A SUPERVISOR

Conference Tree Systems

These feature a new way of looking at and using message systems. While they take a bit of doing and using to catch onto—the results are worth it.

```
    TYPE 'READ HELP' ANY TIME
OR 'READ CONFERENCES' TO START
    COMMAND?   read help complete
    ***HELP                  16-JUL-81
    PARENT = CONFERENCES          USAGE = 848
'READ CONFERENCES' FOR CURRENT SUBJECTS
'READ [NAME]' WHERE [NAME] IS ANY MESSAGE OR SUBMESSAGE.
'BROWSE CONFERENCES COMPLETE' TO SKIM.
'INDEX [NAME]' FOR INDEX OF SUBTREE.
COMMANDS AND OPTIONS MAY BE ABBREVIATED TO THEIR FIRST LETTERS.
PRESS 'S' KEY TO PAUSE OR RESUME PRINTING.
PRESS 'K' KEY TO STOP MESSAGE LISTING.
PRESS 'C' KEY TO KILL CURRENT COMMAND.
CONTROL-H OR DEL KEY FOR BACKSPACE.
TO LEAVE THE SYSTEM, JUST HANG UP!
'READ HELP-COMMANDS' FOR MORE FEATURES.
'READ HELP COMPLETE' FOR ALL HELP DOCUMENTATION.

    THE COMMONLY-USED COMMANDS ARE:
       READ [NAME] [OPTIONS]
       BROWSE [NAME] [OPTIONS]
       INDEX [NAME] [OPTIONS]
       ADDTO [NAME] [OPTIONS]

    THE OTHER COMMANDS, USED LESS OFTEN, ARE:
       TERMINAL
       EXILE [NAME]
       UNEXILE [NAME]
       + ADDTO [NAME]
       SERIAL
```

THE [NAME] ABOVE REFERS TO ANY MESSAGE NAME; FOR EXAMPLE, 'CONFERENCES' OR 'HELP'. [OPTIONS] ARE OPTIONAL SEARCH TECHNIQUES DESCRIBED LATER—THEY CAN BE IGNORED FOR NOW.

THE 'READ' COMMAND PRINTS OUT THE NAMED MESSAGE AND THEN LISTS ITS SUBMESSAGES (CHILDREN) IF ANY—I.E. ALL OTHER MESSAGES

DIRECTLY ATTACHED TO IT. A MESSAGE CAN HAVE ANY NUMBER OF SUBMESSAGES, LIMITED ONLY BY THE SYSTEM'S CAPACITY, AND EACH CAN HAVE CHILDREN OF ITS OWN, TO ALMOST ANY DEPTH. THEREFORE, THE WHOLE STRUCTURE OF MESSAGES FORMS A TREE, WHICH GROWS AS USERS ADD NEW ITEMS TO IT.

THE 'READ' COMMAND ALSO SHOWS THE MESSAGE CREATION DATE, ITS PARENT, AND THE NUMBER OF TIMES IT HAS BEEN READ (ITS USAGE COUNT).

THE 'BROWSE' COMMAND IS LIKE 'READ' EXCEPT THAT IT ONLY PRINTS THE FIRST LINE OF THE MESSAGE. IT IS USUALLY USED WITH SEARCH OPTIONS TO SKIM SECTIONS OF THE MESSAGE TREE.

THE 'INDEX' COMMAND SHOWS THE NAME, CREATION DATE, AND USAGE COUNT OF THE MESSAGE AND ITS COMPLETE SUBTREE (THE NAMES OF ALL ITS CHILDREN AND GRANDCHILDREN) EACH PROPERLY INDENTED ON THE PRINTOUT. IT GIVES A QUICK OVERVIEW OF WHAT IS GOING ON IN A SECTION OF THE TREE OR THE WHOLE TREE. FOR EXAMPLE, TRY 'INDEX HELP' OR 'INDEX CONFERENCES' (YOU CAN ABORT THE COMMAND BY TYPING 'C' OR CONTROL-C WHILE IT IS PRINTING).

'ADDTO' ALLOWS YOU TO ADD A NEW MESSAGE. IT MUST ALWAYS BE ADDED TO A MESSAGE WHICH ALREADY EXISTS IN THE SYSTEM. 'READ HELP-ADDTO' FOR MORE DETAILS.

THE REMAINING COMMANDS ARE USED LESS OFTEN.
'TERMINAL' LETS YOU REQUEST SPECIAL OPTIONS—HALF-DUPLEX, CARRIAGE RETURN NULLS, AND LINEFEED SUPPRESS—WHICH SOME TERMINALS REQUIRE.
'EXILE' AND 'UNEXILE' ARE USED MAINLY BY THE SYSTEM OPERATOR, TO CONTROL OUTDATED OR INAPPROPRIATE MESSAGES WITHOUT DELETING THEM. MESSAGES THAT HAVE BEEN EXILED ARE IGNORED BY 'READ,' 'BROWSE,' AND 'INDEX' UNLESS THEY ARE REQUESTED BY NAME OR UNLESS SPECIAL OPTIONS (DESCRIBED LATER) ARE USED. THE SYSTEM OPERATOR CAN AVOID THE CLUTTER SUCH AS UNWELCOME ADVERTISING MESSAGES, BUT THE READER HAS THE ULTIMATE CHOICE OF WHAT TO SEE—A FORM OF SOFT CENSORSHIP.

***HELP-COMMANDS-2
PARENT = HELP-COMMANDS   USAGE = 160
EVENTUALLY, THE EXILED MESSAGES, AND ANY OTHERS ATTACHED TO THEM, GO AWAY WHEN THE SYSTEM OPERATOR "SQUEEZES" THE DISK TO RECLAIM A SPACE THEY HAD USED. USUALLY, 'EXILE' AND 'UNEXILE' WILL REQUIRE A PASSWORD.

THE ' + ADDTO' COMMAND IS ALSO USED MAINLY BY THE SYSTEM OPERATOR. IT WORKS LIKE 'ADDTO' EXCEPT THAT THE LAST MESSAGE READ OR BROWSED IS RETAINED FOR RE-EDITING. USUALLY, A PASSWORD IS REQUIRED FOR THIS COMMAND, ALSO.

'SERIAL' PRINTS OUT A COPYRIGHT NOTICE AND THE SERIAL NUMBER OF THE PARTICULAR COMMUNITREE SYSTEM. THIS IS PRIMARILY A SOFTWARE PROTECTION FEATURE, ALTHOUGH IT MAY BE HELPFUL IN CASES WHEN ONE CAN'T RECALL WHICH SYSTEM ONE IS CURRENTLY USING.

ALL THE COMMANDS, EXCEPT FOR ' = ADDTO' AND 'SERIAL' CAN BE ABBREVIATED TO THEIR FIRST LETTERS ONLY: R, B, A AND I.

```
***HELP-OPTIONS
PARENT = HELP                          USAGE = 160
```
THE FOLLOWING OPTIONS CAN BE USED WITH THE 'READ', 'BROWSE' AND 'INDEX' COMMANDS:
    COMPLETE
    STARTING [DATE]
    FIND [STRING]
    BEYOND [NAME]
    EXILED
    ONLYEXILED
THESE CAN BE USED IN ANY COMBINATION AND IN ANY ORDER. THEY ARE ENTIRELY OPTIONAL, BUT IF THEY ARE USED, THEY MUST APPEAR AFTER THE [NAME] ARGUMENT OF THE COMMAND.

'COMPLETE' CAUSES THE ENTIRE SUBTREE TO BE PRINTED—ALL SUBMESSAGES AND THEIR CHILDREN, ETC.—NOT JUST THE INDIVIDUAL MESSAGE NAMED. IT IS AUTOMATIC WITH THE 'INDEX' COMMAND, SO IT NEEDN'T BE SPECIFIED. IT IS ALSO AUTOMATIC WITH ALL OTHER OPTIONS, I.E. THEY CAUSE 'COMPLETE' TO BE ASSUMED.

'STARTING' MUST BE GIVEN A DATE IN THE EXACT FORM DD-MMM-YY OR D-MMM-YY, WHERE MMM IS THE FIRST THREE LETTERS OF THE MONTH NAME. 'READ CONFERENCES STARTING 6-JUL-81' WILL SKIP ALL MESSAGES CREATED BEFORE THAT DATE. 'STARTING' IS COMMONLY USED IN EACH SESSION TO CHECK WHAT'S NEW SINCE THE LAST TIME YOU WERE ON THE SYSTEM.

'FIND' MUST BE GIVEN A SEARCH STRING. 'READ GAMES FIND ROLE' WILL PRINT ANY MESSAGE IN A GAMES CONFERENCE (SUBTREE) WITH 'ROLE' IN IT: 'ROLE-PLAYING,' ETC.

A 'FIND' SEARCH OF THE COMPLETE 'CONFERENCES' CAN TAKE A LONG TIME, UP TO 15 MINUTES OR MORE FOR A LARGE DATA BASE. ALL THE OTHER OPTIONS ARE MUCH FASTER BECAUSE THEY USE DATA WHICH IS ALWAYS KEPT IN THE COMPUTER'S MEMORY AND DOES NOT NEED TO BE READ FROM THE DISK. TO MAKE 'FIND' SEARCHES RUN FASTER, SEARCH ONLY THE SUBTREES OF INTEREST, NOT ALL OF 'CONFERENCES,' AND/OR USE 'STARTING' OR OTHER OPTIONS TO LIMIT THE SEARCH. OR USE 'FIND' WITH THE 'INDEX' COMMAND, WHICH IS FAST BECAUSE ONLY TITLES ARE SEARCHED, NOT THE FULL TEXT, AND THE DISK IS NOT INVOLVED. A SEARCH CAN BE INTERRUPTED BY TYPING 'C' OR CONTROL-C.

TO SEARCH FOR A STRING THAT INCLUDES SPACES, ENCLOSE THE
STRING IN DOUBLE QUOTES: 'READ GAMES FIND "ROLE PLAYING" '. IT IS
USUALLY BETTER TO SEARCH FOR SINGLE WORDS OR PARTS OF WORDS
INSTEAD, TO GET SPELLING AND PHRASE VARIATIONS: 'COMPUT' GETS
'COMPUTER,' 'COMPUTING,' ETC.

THE 'BEYOND' OPTION SUPPRESSES PRINTING UNTIL THE NAMED
MESSAGE HAS BEEN SEEN. IT IS USED MAINLY TO CONTINUE A PRINTOUT
WHICH HAS BEEN INTERRUPTED.

'EXILED' INCLUDES MESSAGES WHICH THE SYSTEM OPERATOR OR
OTHERS HAVE EXILED FOR SOME REASON. 'ONLYEXILED' WILL DISPLAY
ONLY THOSE MESSAGES.

ALL OPTIONS CAN BE ABBREVIATED TO THEIR FIRST LETTERS. 'READ
CONFERENCES STARTING 10-JUL-81 FIND GAME EXILED' COULD BE WRIT-
TEN 'R CONFERENCES S 10-JUL-81 F GAME E'. . .

```
***HELP-ADDTO
PARENT = HELP                    USAGE = 168
        ADDTO COMMAND
```

'ADDTO [NAME]' LETS YOU ENTER YOUR OWN MESSAGE TO THE
SYSTEM. YOU WILL BE ASKED TO CONFIRM THE DATE SETTING, AND TO
GIVE YOUR MESSAGE A NAME. TRY TO PICK A NAME THAT IS SOMEWHAT
DESCRIPTIVE OF CONTENT, AND UNLIKELY TO BE USED ELSEWHERE ON
THE TREE, TO AVOID CONFUSION WHEN SEARCHING BY MESSAGE NAME.
YOU MAY ADDTO ANY MESSAGE. ON SOME SYSTEMS, YOU MAY NEED A
PASSWORD TO ADDTO MESSAGES; ON OTHERS, YOU MAY NEED A
PASSWORD TO 'ADDTO PRIVATE' OR 'ADDTO CONFERENCES.' THESE
VARIATIONS DEPEND ON THE SYSTEM OPERATOR AND THE APPLICATION.
YOU CAN EXPERIMENT WITH THE ADDTO COMMAND, AS LONG AS YOU
DON'T USE THE 'SAVEPERMANENT' WHICH WILL WILL SAVE YOUR EXPERI-
MENT PERMANENTLY.

```
***EDITOR
PARENT = HELP—ADDTO              USAGE = 136
        THE EDITOR
```

WHEN YOU USE THE 'ADDTO [NAME]' COMMAND, YOU ARE GIVEN A
SIMPLE LINE-ORIENTED EDITOR WITH WHICH TO ENTER YOUR TEXT. YOU
MAY ENTER UP TO 50 LINES OF TEXT WITH UP TO 80 CHARACTERS IN
EACH LINE. THE EDITOR PROMPTS ARE SELF-DESCRIPTIVE. TO HELP KEEP
THE TEXT READABLE, YOU SHOULD AVOID ENDING LINES WITH PARTS OF
WORDS OR HYPHENATED WORDS. YOU WILL BE ABLE TO 'LIST' AND
'REPLACE' LINES THAT ARE ERRONEOUS BEFORE YOU SAVE THEM
PERMANENTLY. WHEN TYPING A LINE, YOU MAY TYPE THE ENTIRE
80-CHARACTER LINE BEFORE HITTING 'RETURN.' IF YOU GO OVER, THE
SYSTEM WILL GIVE YOU A 'BELL' WARNING. TO GET OUT OF THE EDITOR,
SIMPLY TYPE 'QUIT' WHEN THE EDITOR PROMPT LINE APPEARS, AND THIS
WILL PUT YOU BACK INTO THE MAIN COMMAND MODE.

***MISC-HELP
PARENT = HELP   USAGE = 145
THESE TIPS WILL HELP YOU GET THE MOST OUT OF THE SYSTEM, AND WILL ALSO PROVE BENEFICIAL TO ALL IF THEY ARE HEEDED.

1)   DESCRIPTIVE MESSAGE NAMES, PREFERABLY UNIQUE, ARE RECOM-MENDED. THE USE OF HYPHENS IS PREFERRED OVER THE USE OF PERIODS TO SEPARATE WORDS IN MESSAGE TITLES.
2)   CHILDREN OF EXILED MESSAGES WILL GO AWAY WHEN THE DISK IS SQUEEZED, SO BE CAREFUL NOT TO ADD VALUED MESSAGES TO MESSAGES THAT ARE DATED.
3)   THERE IS A LIMIT OF 321 MESSAGES THAT CAN BE IN THE SYSTEM AT ANY GIVEN TIME, SO ONE BIG MESSAGE IS PREFERRED TO SEVERAL SMALL ONES.
4)   'BROWSE' RETURNS THE FIRST LINE OF A MESSAGE, SO IT IS HELPFUL TO MAKE THAT FIRST LINE AS DESCRIPTIVE AS POSSIBLE OF THE MESSAGE'S CONTENT.

***SYSTEM-PASSWORD
PARENT = HELP   USAGE = 183
THE SYSTEM OPERATOR CAN CHOOSE FROM SEVEN LEVELS OF PASS-WORD PROTECTION—FROM NONE AT ALL, TO REQUIRING A PASSWORD JUST TO LOG ON. AT ANY TIME, THERE IS ONLY ONE PASSWORD, WHICH IS USUALLY GIVE TO ANYONE WHO IS FAIRWITNESS OF A CONFERENCE. THE DEFAULT IS MEDIUM PROTECTION, WITH THE PASSWORD REQUIRED FOR ' = ADDTO', 'EXILE', 'UNEXILE' AND STARTING A NEW PRIVATE CONFERENCE.

A HIGHER PASSWORD LEVEL ALLOWS THE PUBLIC TO READ ONLY AND NOT ENTER ANY NEW MESSAGES; THIS LEVEL MIGHT BE USED FOR A COMMERCIAL MOVIE GUIDE OR SIMILAR SERVICE. IN THIS CASE YOU CAN STILL ADD MESSAGES TO THE 'GUEST' MESSAGE (IF THE SYSTEM OPERATOR HAS PROVIDED ONE), TO GIVE FEEDBACK TO THE SYSTEM OPERATOR ON AN OTHERWISE READ-ONLY SYSTEM.

THE SYSTEM OPERATOR DOES NOT SEE WHAT YOU ARE WRITING OR READING, BUT DOES SEE ERRONEOUS COMMAND NAMES AND ERRONE-OUS PASSWORDS ON A LOG. THIS CAN ALLOW YOU TO COMMUNICATE WITH THE SYSTEM OPERATOR (SAY, TO SEND A PHONE NUMBER TO HIR PRIVATELY. . . SINCE THE PHONE NUMBER IS NOT A VALID COMMAND NAME, IT WILL SHOW UP ON THE LOG). BUT THIS CHANNEL IS UNRELIABLE, AS THE INFORMATION MAY BE DROPPED OFF THE SCREEN BEFORE IT IS SEEN.

Connection-80 Systems

This system, which runs on TRS-80 computers, often attracts a wide cross-section of the telecomputing world.

Commands:

R	...Retrieve a Message	L	...Leave a Message
S	...Scan Messages	T	...Terminate Session
K	...Kill a Message	I	...System Information
U	...User Log Display	B	...Bulletin Display
E	...Elapsed Time on System	C	...Chat with SYSOP
M	...Merchandise Review	P	...Purchase Merchandise
X	...Expert User Mode	D	...Download Section
H	...Help (This Menu)	O	...Return to Logon
A	...Alter Baud Rate	F	...Format Screen

Forum-80 Systems

Another system which runs on TRS-80 computers but welcomes all.

S = Summarize messages
R = Read messages in system
E = Enter messages
K = Kill messages
F = Flagged msg retrieval
B = Bulletins
I = Information about system
H = Help with sys operation
U = User log
C = Configuration changes
T = Terminate Connection (SIGN OFF)
O = Other BBS Systems
L = Local features (Editorials, New User Info, Download, Upload, Password Info)

03:45 COMMAND (Hit ENTER for list):**H**

(H) HELP

S = Summary		R = Reading messages
E = Enter messages		F = Flagged retrieval
M = Multiple commands		C = Control characters/codes
A = Abort to COMMAND mode		

HELP WITH: SUMMARY

The message SUMMARY subfunction is entered from COMMAND mode with command 'S'.

SUMMARY SUBCMD functions are:

   C — Complete summary. Prints all information contained in the message header.

Q — Quick summary. Prints an abbreviated summary consisting of only the msg/status line and the subject line.

S — Search summary file. Allows the user to search the summary file and see a summary of all the messages on file which match the search criteria. String searches may be conducted on the following fields:

>   FROM:
>   TO:
>   SUBJ:

Searches may also be conducted for all messages within a given category. The present message categories are:

>   MISC
>   PERSONAL
>   COMMERCIAL
>   SYSTEM BULLETINS

The complete summary and quick summary may be operated in both forward and reverse. The starting point may be specified by the user by entering a starting message number. Hitting a carriage return for starting message will default as follows:

> FORWARD: Starts at oldest message since your last call. (First message in file for new callers)
>
> REVERSE: Starts at newest message.

Messages may be flagged in complete, quick and search functions. See flagged message help for details.

HELP WITH: ENTERING MESSAGES

The message ENTRY subcmd list allows message entry in the following categories:

> Miscellaneous — Any message not falling into one of the following categories.
>
> Personal — Password protected messages of a personal or confidential nature.
>
> Commercial — Messages containing advertising of goods or services for sale or wanted to buy.

Messages may be entered into the system by two methods.

Entry modes:

Line entry   — Standard line oriented manual entry of text lines up to 80 characters per line and up to 26 lines per message. Maximum total characters per message 1024. In line entry mode the system prompts with:

LINE NO.   (NO of CHAR LEFT)

To exit line entry mode before the 26th line answer the prompt with a carriage return as the first character of the line.

Block entry — Used for auto entry purposes only. The message text is transmitted to the system as one continuous stream of characters. For best results, your configuration should be to 8 bit words, 2 stop bit and PARITY DISABLED! If the text transmission is less than 1024 characters, the input mode is exited by transmitting E.O.T. (04H). No editing or continue functions available for block entry.

Passwords:

The user is asked to specify a password for every message. This password may be from 1 to 8 characters. For personal messages this password becomes the retrieval password. For all messages this password is required to kill the message.

Saving your message:

After you have entered your message and are satisfied it is correct, you may save it in one of two ways:

Save — The save ('S') subcmd saves your message to disk and returns to the message entry subcmd prompt. You may then enter another message or abort back to command mode.

Terminate then save — The 'T' subcmd terminates your phone connection then saves the message to disk. If you are finished using the system, you can save a few seconds of phone time by using this subcmd.

Message attributes:

Certain attributes may be assigned to your message. They are:

R — Auto-kill after message is (R)etrieved by the person to whom it is addressed. Be careful to spell the name exactly the way the user is logged or this function will not activate.

D — AUTO-KILL on a specific date. This is used to limit the time the message remains in the system to less than a month. When prompted by the system, enter the DAY ONLY. For example, if you are placing a message on April 7 and you wish it to remain 2 weeks, just enter 21 as the AUTO-KILL date. If you entered 21 as the AUTO-KILL date and you were placing the message on the 22nd, it would not AUTO-KILL until the 21st of the following month.

Be careful to enter a legitimate date. If you enter 31 and the month only has 30 days, the message will remain until it dies naturally or until a month with 31 days is reached.

HELP WITH: RETRIEVAL

The message RETRIEVAL subfunction is entered from COMMAND mode with command 'R'.

RETRIEVE SUBCMD functions are:

F — Forward sequential retrieval. Operates the same as the forward complete summary except that the message text is also printed.

R — Reverse sequential retrieval. Operates the same as the reverse complete summary except that the message text is also printed.

P — Personal message retrieval. Retrieves personal messages only if the user enters the correct password. Personal messages are flagged for auto-kill. You may view your personal message as many times as you wish but as soon as you terminate the system will kill it. Personal messages remain in the file only until viewed by the intended recipient. The auto-kill of these messages releases disk space immediately for the use of the open message file.

I — Individual message retrieval. Retrieves messages individually by number.

S — Search message file. Operates the same as the search summary subfunction except it also prints the message text.

HELP WITH: FLAGGED RETRIEVAL

Messages are flagged as follows:

1. Enter the summary information of your choice. (Complete, Quick, Search in either forward or reverse.)

2. As each summary header is printing, decide if you wish to view that message.

3. If you wish to view the message, wait until the header has finished printing, then hit 'F' any time during the printing of the following

header. As each message is flagged you will see a flag prompt with the message number flagged. In order to be able to flag the most recent message, you should flag the SUMMARY in the REVERSE direction.

4. Execute flagged retrieval with command 'F'. The flagged routine is also accessible from the SUMMARY subcmd with subcmd 'F'.

5. If the message alert feature flagged any messages for you at logon they will be retrieved along with the messages you flag manually.

The flagged message function can be quite effective when used in conjunction with the summary search. When the summary prints those messages that match your search criteria, you can further refine the search by flagging just the ones of that group that you wish to see.

HELP WITH: MULTIPLE COMMANDS

The system is equipped to receive mutiple commands in a single input. Multiple commands ARE ACCEPTED ONLY IN THE COMMAND MODE INPUT! SUBCMD PROMPTS DO NOT ACCEPT MULTIPLE COMMANDS. The multiple command string may consist of from 1 to 20 sequenced commands AS LONG AS ALL COMMANDS OPERATE WITHIN THE SAME SUBCMD FUNCTION! The multiple command sequence is terminated when you re-enter command mode.

The choice of the delimiter between commands is up to the user. Any non-alphanumeric ASCII character may be used but the same delimiter must be used throughout the multiple command string. For example, if you are inputting a MULT/CMD string which will not include a search string, then a space may be used as a delimiter. However, if your MULT/CMD string will include a search string with an imbedded space, such as 'JIM BROWN', then another delimiter such as , or ; or : should be used.

It is strongly recommended that you form a habit of using either colon (:) or semicolon (;) as your delimiter as changes in the MULT/CMD function in the near future may necessitate restricting the allowed delimiters to these characters.

Below are some examples of useful mult/cmd strings:

00:00 COMMAND: R,F,D,A,

1. Retrieve
2. Forward
3. Default to beginning message (a msg number may be substituted here)
4. Abort back to command mode

00:00 COMMAND: R,S,D,F,JIM BROWN,Y,A,

1. Retrieve
2. Search message file
3. Default (start at bottom msg for search of entire file)
4. Search the FROM field
5. Search for the string 'JIM BROWN'
6. Yes (this skips the verification question)
7. Abort back to command mode when finished

00:00 COMMAND: O,S, 714-,S, TX ,A

1. Other systems phone numbers
2. Search file
3. Search for all systems in '714' area code
4. Search file (again)
5. Search for all systems in 'TEXAS'
6. Abort back to command mode when finished

The MULT/CMD input has also been extended to the 'FIRST NAME' question at log on. This will allow the user to input a single string and skip the remaining questions. This will be particularly useful to users with smart terminal software having auto log features.

Example of single string log on:

YOUR FIRST NAME ? JIM;BROWN;NEW YORK, N.Y.

When entered as above, the system will echo back the name and city and ask for verification. To skip the verification questions, add ;Y to the end of the string. To skip the bulletins, add ;S after the ;Y.

The log on string will not accept all the delimiters accepted by command mode. Since / , . - (and space) may appear in the string, do not use these as delimiters. Again it is that ; or : be used as delimiters.

These are just a few simple examples of what the MULT/CMD feature will do. With a little practice, I'm sure you will find this a great help in efficient system use.

HELP WITH: CONTROL CHARACTERS/CODES

Certain system control functions are executed by control characters/codes.

Control characters:

S — 'STOP' Stops the current function and returns control to the command or subcmd mode.

P — 'PAUSE' Pauses and allows 2 minutes to study the display. Hit any character to continue.

N — 'NEXT' Used during multiple message retrievals. Skips the message currently printing and goes to the next message in sequence.

F — 'FLAG' Used during summary to flag messages for flagged retrieval. (See HELP: Flagged retrieval)

The above are alphabetical keyboard characters, NOT CONTROL CODES!

S & P respond instantly during the printing of message text and at the end of the line elsewhere in the system.

Control codes:

Certain control codes are recognized by the system during data input by the user. These codes are as follows:

Control 'H' — (08 Hex) Backspace. If your terminal backspace key transmits a rubout (7F Hex) you must redefine your key or use a control 'H' to perform a backspace. The system configuration table may not be used to redefine backspace at this time.

Control 'I' — (09 Hex) Tab. Tabs, input. Tab stops are at 0, 8, 16, 24, 32, 40, 48 & 56. Tab should be used only with message text input as may cause unpredictable results with mult/cmd input.

Control 'X' — (18 Hex) Delete. Deletes the entire line currently in the input buffer.

TRS-80 users may obtain the above codes with:

Left arrow (backspace)
Right arrow (tab)
Shifted left arrow (delete)

Net-Works Systems

Runs on the Apple II computer, but supports all callers.

HERE IS LIST OF SUPPORTED CATEGORIES:

(A) PPLE BULLETIN BOARD SYSTEMS. ALSO INCLUDES SOME NON-APPLE SYSTEMS.
(B) ULLETIN BOARD (READ/LEAVE MESSAGES
(C) HAT WITH SYSTEM OPERATOR
(D) OWNLOAD PROGRAMS (APPLE, PET, TRS-80)
(G) OODBYE. ACCEPTS USER FEEDBACK BEFORE HANGING UP.

(H)ELP WITH THE SYSTEM. A DETAILED LIST OF SYSTEM FUNCTIONS AND SUB-FUNCTIONS.

(I) NFORMATION ABOUT THIS SYSTEM

(M)AIL. SEND/RECEIVE PRIVATE LETTERS.

(N)ULLS ON OR OFF FOR PRINTER/TERMINAL

(O)FF. CAN BE USED ANYTIME. THIS WILL ALLOW IMMEDIATE SIGN-OFF.

(P) ROGRAMMING TIPS FOR APPLE USERS.

(S) PECIAL INTEREST TOPICS.

(T) IME. TURNS ON/OFF DISPLAY OF CONNECT TIME. ALSO DISPLAYS CURRENT TIME.

(U) SERS LISTING OF SYSTEM CALLERS

(?) DISPLAYS THIS MENU.

=:=

ENTER (A,B,C,D,G,H,I,M,N,P,S,T,U,?): **H**

=:=

**** SYSTEM FUNCTIONS EXPLAINED ***
:::CTRL-S = STOP :: CTRL-Q = START :::
(ANY OTHER KEY TO QUIT DISPLAY)

FROM MAIN MENU

A = DISPLAYS OTHER APPLE BULLETIN BOARD SYSTEMS, INCLUDING THOSE THAT ARE SIMILAR TO THIS ONE (GROUPED TOGETHER AT THE TOP OF THE LIST AND PRECEEDED BY "***"). ALSO INCLUDES SOME SYSTEMS THAT ARE NOT APPLE-BASED. THE LISTING IS MAINTAINED IN ALPHABETICAL ORDER.

B = BULLETIN BOARD ACCESS. THIS IS A PUBLIC BOARD THAT ALLOWS READING AND POSTING OF MESSAGES. USUALLY USED TO EXCHANGE INFORMATION OF A GENERAL NATURE (SUCH AS ITEMS FOR SALE, HELP WITH A PROBLEM, ETC.). WHEN POSTING MESSAGES THE SYSTEM AUTOMATICALLY ADDS YOUR NAME AND DATE & TIME MESSAGE POSTED. ONCE YOU'VE ACCESSED THE BULLETIN BOARD THERE ARE JUST FOUR SIMPLE COMMANDS FOR OPERATING IT:

   S = SCAN MESSAGE TITLES AND READ MESSAGES.

   L = LEAVE MESSAGES (ONLY IF YOUR PASSWORD HAS BEEN VALIDATED

   D = DELETE MESSAGE (ONLY IF YOUR PASSWORD HAS BEEN VALIDATED)

   Q = QUIT THE BOARD. RETURNS YOU TO THE MAIN SYSTEM.

C = ALLOWS A 2-WAY CONVERSATION WITH THE SYSTEM OPERATOR IF HE OR SHE IS AVAILABLE. THIS IS MOST USEFUL IF YOU HAVE ANY QUESTIONS OR PROBLEMS THAT NEED TO BE MENTIONED. THE

CONVERSATION TAKES PLACE BY TYPING, NOT TALKING, BACK &
FORTH. AFTER YOU HAVE SELECTED THE (C)HAT OPTION YOU CAN
CONTINUE USING THE SYSTEM (RATHER THAN WAITING FOR THE
OPERATOR TO ANSWER). IF THE OPERATOR IS AVAILABLE HE OR
SHE WILL BE AWARE THAT YOU WANT TO CHAT AND CAN ANSWER
YOU NO MATTER WHERE YOU HAPPEN TO BE IN THE PROGRAM.

D = DOWNLOADING OF PROGRAMS. THE SYSTEM HAS THE ABILITY TO
SEND YOU MANY KINDS OF PROGRAMS FOR THE MOST POPULAR
COMPUTERS. UP TO 9 CHOICES MAY BE ON THE SELECTION LIST AT
ANY ONE TIME. IF YOU ARE USING AN APPLE WITH AT LEAST ONE
DISK DRIVE AND A HAYES MICROMODEM THEN THIS SYSTEM CAN
SEND AND SAVE YOUR PROGRAMS AUTOMATICALLY. IF YOU ARE US-
ING ANY OTHER CONFIGURATION THEN THE SYSTEM ASSUMES YOU
HAVE A METHOD OF CAPTURING THE DATA AS IT COMES ACROSS
THE PHONE LINE AND WILL SEND (NOT SAVE) YOUR PROGRAM.

G = GOODBYE. THIS IS ONE OF TWO WAYS TO PROPERLY SIGN-OFF THIS
SYSTEM (THE 2ND WILL BE EXPLAINED LATER). BEFORE LEAVING
THE SYSTEM YOU ARE ASKED FOR ANY COMMENTS YOU MIGHT
WISH TO MAKE. TO EXIT FROM THE COMMENTS (WE CALL IT FEED-
BACK), ALL YOU HAVE TO DO IS TYPE THE THREE CHARACTERS: /EX
(BUT DON'T WORRY ABOUT REMEMBERING THIS, YOU WILL BE
REMINDED THEN WHEN THEY ARE NEEDED).

H = SYSTEM INFORMATION. THIS CONTAINS INFORMATION THE SYSTEM
OPERATOR WANTS YOU TO KNOW ABOUT. USUALLY, WHAT TYPE OF
SYSTEM IS BEING USED, HOURS OF OPERATION, ETC.

M = ELECTRONIC MAIL. A PERFECT WAY TO COMMUNICATE WITH
ANOTHER PERSON PRIVATELY. NO ONE ELSE EXCEPT THE PERSON
THE LETTER WAS ADDRESSED TO WILL SEE YOUR MESSAGE. IF YOU
HAVE MAIL WAITING, THE SYSTEM WILL NOTIFY YOU IMMEDIATELY
AFTER YOU SIGN-ON, SO THERE'S NO NEED TO SEARCH ANY
SPECIAL AREAS. YOU WOULD SELECT THE M-OPTION TO READ MAIL
THAT THE SYSTEM NOTIFIED YOU ABOUT OR TO LEAVE MAIL FOR
SOMEONE ELSE.

N = NULLS. THIS IS WHAT TO USE IF YOU WANT TO SLOW DOWN THE
SPEED THAT THE SYSTEM IS SENDING AT. NULLS IS ANOTHER WAY
OF SAYING THAT THE SYSTEM WILL PLACE A DELAY AT THE BEGINN-
ING OF EACH LINE IT WILL SEND. THE DELAY TIME CAN BE UP TO A
MAXIMUM OF 2-½ SECONDS. IF YOU ARE USING A PRINTER AND IT IS
MISSING CHARACTERS AT THE BEGINNING OF EACH LINE THEN
YOU'LL WANT TO SELECT THIS OPTION.

O = OFF. THIS IS THE OTHER WAY TO SIGN-OFF THIS SYSTEM. UNLIKE
THE G-OPTION, THIS ONE ALLOWS IMMEDIATE SIGN-OFF WITHOUT
COMMENTS OR FEEDBACK. ALTHOUGH THE OPTION APPEARS ONLY
AT THE MAIN SYSTEM SELECTION LIST, IT IS ACTUALLY AVAILABLE

FROM JUST ABOUT ANYWHERE IN THE SYSTEM. YOU CAN TYPE "OFF" IN PLACE OF SOMETHING ELSE FOR VERY FAST SIGN-OFF.

P = PROGRAMMING TIPS. AVAILABLE TO USERS WITH VALIDATED PASSWORDS. YOU'LL FIND MANY INFORMATIVE ITEMS TO HELP YOU USE YOUR APPLE. NEW ITEMS ARE ALWAYS PLACED AT THE TOP OF THE LIST SO YOU DON'T HAVE TO LIST EVERYTHING JUST TO FIND OUT WHAT'S NEW. CHECK THIS SECTION OFTEN, ESPECIALLY IF YOU LIKE TO PROGRAM (FOR FUN OR PROFIT).

S = SPECIAL TOPICS. EIGHT CATEGORIES THAT CAN CONTAIN JUST ABOUT ANYTHING. SUBJECT TITLES MAY CHANGE FROM TIME TO TIME. CATEGORIES THAT CURRENTLY CONTAIN INFORMATION ARE PRECEDED BY AN ASTERISK (*). USUALLY, THE SUBJECT MATTER HERE IS MORE DETAILED AND LONGER THAN WHAT COULD NOR-MALLY BE POSTED ON THE BULLETIN BOARD.

T = TIME. IF YOU WOULD LIKE THE SYSTEM TO HELP YOU KEEP TRACK OF HOW LONG YOU'VE BEEN ON THE SYSTEM THEN THIS OPTION WILL DO IT. IF AT ANY TIME YOU WANT TO TURN OFF THE DISPLAY, JUST SELECT THIS FUNCTION AGAIN. THINK OF IT AS AN ON-OFF SWITCH.

U = USERS LISTING. THIS LIST IS MAINTAINED IN ALPHABETICAL ORDER BY LAST NAME AND IS AVAILABLE TO ALL THOSE WITH VALIDATED PASSWORDS. YOU CAN LIST ALL THE USERS OR JUST A SELECTED GROUP. IDEAL FOR SEEING IF YOUR FRIEND ALSO CALLS THIS SYSTEM OR FOR GETTING THE CORRECT SPELLING OF A NAME TO SEND MAIL TO OR JUST FOR BEING CURIOUS.

? = THE QUESTION MARK WILL DISPLAY THE CURRENT SELECTION LIST. USE THIS IN CASE YOU'RE NOT SURE OF WHICH OPTION YOU WANT.

GENERAL COMMENTS

THROUGHOUT MOST OF THIS SYSTEM A CTRL-S WILL TEMPORARILY STOP THE SYSTEM AND A CTRL-Q WILL RESUME THE DISPLAY. PRESSING ANY OTHER KEY WILL END THE DISPLAY AND PROMPT YOU FOR WHAT ACTION YOU WANT TO TAKE NEXT.

WHEN TYPING FROM YOUR KEYBOARD, THE FOLLOWING COMMANDS ARE AVAILABLE:

CTRL-W = BACK UP ONE WORD.
CTRL-X = BACK UP ONE LINE.
CTRL-H = BACK UP ONE CHARACTER.
CTRL-R = RETYPE CURRENT LINE.

NOTE THAT CTRL-U IS THE SAME AS CTRL-X AND THAT THE RUBOUT KEY WILL DO THE SAME THING AS CTRL-H. THAT MEANS THAT ON THE APPLE,

THE LEFT ARROW WILL BACK YOU UP ONE CHARACTER AND THE RIGHT ARROW WILL ERASE AN ENTIRE LINE.

TO END ANY MESSAGE YOU ARE LEAVING ON THIS SYSTEM, USE THE 3 CHARACTERS: /EX (FOR EXIT) ON A NEW LINE ALL BY THEMSELVES AND PRESS THE RETURN KEY. THESE CHARACTERS WILL NOT BE PART OF YOUR MESSAGE.

THE SYSTEM CONVERTS ALL CHARACTERS RECEIVED TO UPPER CASE SO THAT ALL MAY BE ABLE TO USE THE SYSTEM.

YOU MUST PRESS THE "RETURN" KEY AT THE END OF EVERY LINE YOU TYPE. THE SYSTEM WILL NOT ALLOW YOU TO CONTINUE UNLESS THIS IS DONE. YOU WILL ALSO HEAR A WARNING BELL 5 CHARACTERS BEFORE THE END OF EACH LINE AND A BELL FOR EACH CHARACTER AFTER YOU HAVE REACHED THE END OF THAT LINE. WHEN THIS HAPPENS, IT IS YOUR SIGNAL TO PRESS THE "RETURN" KEY TO ADVANCE TO THE NEXT LINE.

PMS Systems

Here is the complete help file for the Public Message System software which runs on an Apple II computer but will accept calls from any computer or terminal. If you are following along in our "Your First Phone Call" chapter—this is the style of bulletin-board you should be calling.

Command ?**H;ALL**

Format for printer (y/n) ?**N**

* * System Control functions and codes * *

There are certain characters which cause specific functions to occur while system is printing to you. (NOT when waiting for input.) These may be control, lower or upper case characters.

— C — Stops printing the current line up to carriage return. Use this if you want to skip over several lines of text without aborting the function. Use one 'C' for each line you want to jump over.

— K — Causes a jump to the next logical operation. As an example, if you were retrieving several messages, it would cause a jump to the next message. During message entry listing, it will return you to the message entry command level. The ONLY time this command will return to main COMMAND level is if there is no logical next function. It can also be used to cancel the bulletins at sign-on, and jump directly to the message alert routine. (Two K's during bulletins would cancel bulletins and message alert and go right to COMMAND.

— N — Adds two nulls for each N typed. Use this when you have a printer online, which needs nulls (to slow down a Scan or Quickscan if you are having trouble keeping up with it. See also 'P'.)

— N —   resets nulls to zero. Regardless of current null setting, resets null to zero. You can then add again with N if you wish.

— R —   Flags messages during a scan or quickscan to be later retrieved with the * command. Type an R at the NEXT message header. It will always be one message behind. There is usually not time to comprehend a message header and type the R before that header is done being printed. So just pause slightly and then type it. At the end or that next header, you will get verification and number of the message flagged.

— S —   Stops text output until any other character is sent.

— X —   This causes an unconditional abort of whatever function you're in, and an immediate jump to main COMMAND level.

* * System Commands * *

All system commands are input to the main COMMAND level. Certain commands, H,K,R and S may also have extenders describing or anticipating the next question asked. As an example: Since H is for help, if you wanted help on everything, you could type H;ALL. Using R as another example, if you wanted to retrieve message –1937 you could type R; 1937 etc. More examples of various multiple parameters are explained for each individual command.

— B —   Reprints the bulletins that are displayed when you first sign on to the system.

— E —   Enter a message into the system. Pretty self explanatory, just follow prompting. You must enter a password when asked. This is the password used to kill the message. There are also two levels of security messages available. If you type LOCK as your password (you will be asked again for the password for killing), the message will be marked as private, and will automatically open for the person to whom its addressed or to who wrote it. Others will be asked for a password which will be the same as the password for killing it. If you enter LOCX instead of LOCK, the message can be read ONLY with the password.

— G —   Goodbye. Exit the system and hangup the phone. Files will be updated at this time. System will also respond to: END, OFF, BYE and a few others.

— H —   System help files. Typing an H by itself will print out all the possible areas you obtain help for. Typing H;(character) will print help on a specific function.

— K —   Kill a message. This will remove a message from the system files. You must have the password (entered during message entry) to use this. You also have the option of automatically killing messages that are addressed to you at the time you read them. See more details in R command.

— M — Message alert. This command is issued automatically directly after the bulletins when you sign on. It allows you to automatically retrieve all messages addressed to you. You may use it at anytime in the program. One word of caution: If you have flagged messages for retrieval and use this command all the flagged messages will be lost. If you do use it, and did not retrieve all the messages to you, you can continue with the * command. (Same as flagged retrieval)

— O — Other systems list. Updated regularly, contains a summary of all known public access message systems of all types, in alphabetical order. Can also be printer formatted.

— Q — Quickscan of message headers. See also, S.

— R — Retrieve a message from the files. There are several modes here. You can select messages singly, or in multiples. Examples of entries: R;381 or R381 or R381;560;etc. To retrieve all messages starting from a certain point, type: R;555   or R;1010$. See also the * command for flagged msgs. When you retrieve a message that is addressed to you, at the end of it, you'll be asked if you want to kill it (automatically), and then if you want to reply to it. The automatic reply does some of the busywork for you (To: From: etc.) and upon completion of the reply and saving the message, will continue with retrieving other messages you may have specified. During auto-reply, if you just type carriage return in response to Subject?, it will take the subject of the message that was to you and add a R/ to it, meaning reply to:

— S — Scan message headers. Here you can specify a starting message number to start the scan. S;500 would scan starting at message 500 or the next highest if there is no message 500. You can also scan in reverse order by either specifying a number   [5] the highest message number, or by adding a '-' (minus) sign directly after the message number.
Examples: S;500 S1200 S;1040- etc. As you are scanning you use 'R' to flag messages you want to read later with the * command. See *.

— SR — Selective retrieval. Use this command to retrieve all messages whose headers contain data you are looking for. As an example, if you entered FOR SALE, it would automatically retrieve all messages with FOR SALE in the header. This will work for ALL aspects of the header. FROM, TO, DATE, SUBJECT and LOCATION. Even number of lines. All messages that meet these parameters are put in flagged memory and automatically retrieved. If there are not matches, it will tell you.

— T — Prints current Pacific coast time and date, and the length of time since you logged on.

— U — User modified functions. These are parameters which affect certain default conditions of the system. You modify them to your current needs.

A — Apple 40 mode. Toggles between 40 and 60 column message entry mode. Does not affect text output.

C — Case switch. Toggles between upper only and upper/lower text OUTPUT. Lower case text is accepted during message entry and comments in either mode.

D — Duplex switch. Toggles between full (echo) and half (no echo) duplex modes. At this point in time, this function is only supported in Printer terminal mode (V) modifier.

L — Line feed switch. Turns linefeeds on or off. System default when you call, is on.

N — Nulls. Displays the current number of nulls in effect, and allows you to modify them directly.

P — Prompt. Allows you to change the current system prompt, which is usually a (?) question mark. Here you can enter either the character you want the prompt to be, or the Ascii value of it. The prompt will stay defined as such through all system functions, until you change it again, or hang-up. This feature can be used to good advantage with automatic upload or download programs. (automatic message entry, etc.)

V — Video/printer terminal mode toggle. In the video mode will accept data at full 300 baud. Recognizes ctrl-h underscore and rubout (asc08, 95 and 127 respectively) as backspace characters. In the printer mode, same characters are recognized, but during the backspacing additional reverse slashes and the characters being backspaced over in reverse order will appear. This mode should not be used for automatic upload programs, as it is quite slow. Hope to change it soon.

STAT — Displays current system status of all the modes covered by the U function, so you can see your current setup.

— X — eXpert user mode. Does away with the prompting at command level, the pausing between messages during a multiple retrieval and allows certain other priveledges.

— Z — Allows you to resume entering a message after you have aborted it. Let's say you were entering a message and realized you had forgotten something relevant to that message, and needed to reread an earlier message or article. You can abort the message, reread other messages or articles and then reenter your message with all data intact, and continue. You cannot however, kill a message and then return.

— ? —   Prints a list of all available commands at your current mode of operation.

— * —   Retrieve messages in flagged memory. This can be messages flagged during a scan or using the M or SR function. You may pick up where you left off with this command, if you ended your retrieval for any reason. Using this command with scanning messages, (such as S;1000*) will clear the flagged memory completely for starting fresh.

* * Message entry commands * *

— A —   Abort message and return to command level. You may continue your message with the Z command if you have not tried to kill another message.

— C —   Continue with message entry. Allows you to continue your message at whatever the next line in succession would be.

— D —   Delete a line. Specify the number of the line you wish to delete. D;x & Dx are legal here. (x = line number)

— E —   Edit or retype a message line. To replace the line, specifying a string you want to replace with what you want to replace it with, in this form: OLD/NEW with old and new string separated by a slash. You can also use three slashes as in the old editing routine: /OLD/NEW/. To remove a section of text type: OLD// or /OLD//. To insert text at the beginning of a line type: this is new text. The left arrow means insert at the beginning. To append text at the end of a line type: this is new text. The right arrow means append to the end. If you type a control-s (control-s only) as the first character of the line, the contents of the line will be automatically centered. If you are modifying an existing line and just type a control-s by itself, whatever is already on that line will be centered. Control-s as the first character of a line will also work during regular line entry (not just during edit). Ex and E;x (x = line number) are legal here.

— I —   Insert a line. (Ix or I;x also legal) Allows you to insert a line directly BEFORE whatever line – you specify.

— S —   Save the completed masterpiece to disk. You MUST use this command for your message to be saved!

— W —   reWrite an old message. Using this command, you can kill an older message (with the correct password, or course) with the contents of it appearing in the message you are currently entering. The old message lines will be appended to your current position in the message. As an example, if you were at line 10 of a message, and wanted some data that was in an old message, use W and when it comes back, the contents of the message you killed will start at line 10. You can then edit or modify as required, then save it back out.

RCP/M Systems

These Remote CP/M systems may be running on just about any style of micro or minicomputer which operates CP/M. While any computer can call-in to read bulletins and leave messages most of these systems allow another CP/M to take full control of their disk drives for downloading their many pubic domain programs.

Welcome to my remote Z-80 CP/M system. RIBBS is available for leaving messages. Works like the big CBBS system minus a few features. Give it a try. Just type:

```
A   USER 3
A   RIBBS P
```

and you will be in the program. Option C will return you to CP/M. Your messages are automatically saved on disk. Please kill messages addressed to you after you have read them. This will help conserve disk space.

On this system you now are the CONSOLE. The LPT (Line printer) is a null device that goes nowhere.

For more information on how this system is configured and where various files are kept, type:

```
A   TYPE THIS-SYS.DOC
```

If you would like a detailed description of how CP/M (the DOS you are now using) operates, type:

```
A   INFO
```

Have fun! When you are done, do this:

```
A   BYE
```

The system will automatically terminate upon loss of carrier. If you have any problems with the system, you can reset it by hanging up and calling back again.

```
A   type this-sys.doc
```

MLIST ver 5.0—Multiple file lister
CTL-S pauses, CTL-X skips to next file, CTL-C aborts

    LISTING FILE: THIS-SYS.DOC

Organization of the disks—

Drive A contains system utilities which are (mostly) not copyable. The CCP is patched to automatically search drive A if logged into B or C and a .CON file is not found. (See CCPPATCH.ASM on drive B)

A   TYPE SYSTEM.NOS for a list of other remote systems' phone numbers.
A   TYPE OTHERSYS.NOS for additional remote system listings.
A   TYPE RCPMLIST.* for numbers and information and Remote CP/M systems.

Drives B and C contain available programs, categorized by user number as follows:

    user 0—Assembly-language
    user 1—PL/I and C
    user 2—FORTRAN
    user 3—BASIC

Use XMODEM to transfer programs to/from your system. A   TYPE XMODEM.DOC for details on use.

Index

NOTES

NOTES